DICTIONARY OF
SCIENCE

GEDDES & GROSSET

DICTIONARY OF
SCIENCE

Published 2001 by Geddes & Grosset, an imprint of
Children's Leisure Products Limited

© 1998 Children's Leisure Products Limited,
David Dale House, New Lanark ML11 9DJ, Scotland

First published 1998
Reprinted 2001 (twice)

ISBN 1 85534 346 0

Printed and bound in the UK

A

aberration is what happens in physics when a lens (or curved mirror) does not produce a true image. It is caused by the light rays from any point on the object being viewed forming an image (focusing) at slightly different positions. In astronomy, aberration refers to the apparent change in position of a star. This is due to the fixed speed of light travelling from the star and the movement of the person looking at the star, because the Earth is moving in its orbit.

abscissa the term given to the horizontal or x-co-ordinate when describing the position of a point on a graph. It is the distance of the point from the vertical or y-axis. For example, a point two units along the x-axis and three units up the y-axis is said to have CARTESIAN CO-ORDINATES (2, 3) where the abscissa is 2.

absolute temperature a temperature measured on the KELVIN SCALE with respect to ABSOLUTE ZERO.

absolute zero the temperature at which the particles that make up matter have no energy at all whether due to heat or motion. It is theoretically given the value of -273.15°Celsius (-459.67° Fahrenheit).

ABO blood group see **blood grouping**.

AC see **alternating current**.

acceleration is when the speed at which an object is travelling is further increased. The object travels faster because some force is applied to push or pull the object, making it move more quickly. In physics, when calculations are made concerning acceleration, it is usually represented by the symbol a and is measured in metres per second, per second or metres per second squared (written in mathematical shorthand as ms^{-2}).

When bodies fall freely under the influence of the force of GRAVITY they all fall at the same rate, that is at a uniform acceleration. Whether the object is a lead weight or a feather they fall at the same rate due to gravity. The reason that in reality the lead weight hits the ground before the feather is due to air resistance holding back the feather. Acceleration due to gravity is approximately 9.8 metres per second per second so when the air resistance is very little, the velocity (speed) of a falling object increases by 9.8 metres per second, per second. On the other hand, an object shot straight upwards with a particular velocity will slow down (decelerate) by the same rate every second until it reaches its greatest height after a particular period of time. When this maximum height is reached, the object that had been shot straight upwards (no horizontal motion whatsoever) will start to fall with an acceleration of 9.8 ms^{-2}.

acetic acid a chemical compound, more correctly called ethanoic acid, which is the acid found in vinegar. It is present as a solution in water of strength 3 to 6%, in other words there are 3 to 6 parts of acetic acid in 100 parts of water. The chemical formula for acetic acid is CH_3COOH which contains two carbon atoms bonded together; to one are bonded three hydrogen atoms and to the other one oxygen and one hydroxyl group (OH). Acetic acid is made from ethanol or acetylene.

acetone a chemical compound known also as propanone (or dimethyl ketone CH_3COCH_3) which is an important SOLVENT used a great deal in industry e.g. in making plastics.

acid are chemical substances that occur in many forms and have numerous uses. Although some acids are harmless, there are many which can cause burns to the skin and some are poisonous and corrosive. Some are essential to life and, for example, are found in our stomach where they help to digest food.

Relatively harmless acid (citric acid) occurs in fruit such as oranges, lemons and grapefruit. Formic (or methanoic) acid is found in stinging nettles while in industry many strong acids (hydrochloric, nitric and sulphuric acids) are used in commercial manufacturing processes. In the chemistry laboratory a whole range of acids is used in chemical experiments. The acidity of an acid can be shown by the pH of the solution and is determined by the number of hydrogen ions in the solution. An acid is therefore a substance which releases hydrogen ions during a chemical reaction. An acidic solution has a pH of less than 7 and will react with a BASE to form a salt and water. When a base and acid react together in the correct proportions, the two substances balance each other, or neutralize each other. The pH of a solution is measured using indicators, one of which is litmus paper. Acids turn litmus paper red.

acid rain gases, known as industrial pollution, given off by car exhausts, power stations burning coal and factories rise into the atmosphere and react with rain and since the gases are acidic in nature, they form dilute acid, or acid rain. The main gases are oxides of nitrogen and sulphur dioxide and because they can travel great distances in the air, the resulting acid rain may fall many miles from the source of pollution - even in a different country. It then has harmful effects on both plant and animal life either by scorching leaves or stripping trees or by making lake waters so acidic that life cannot be maintained. Efforts are now being made to reduce the gas emissions from industrialized countries.

acid rocks a class of igneous rocks in which there is over 66% SILICA by weight. Most of the silica is in the form of silicate minerals but the extra forms free QUARTZ within the rock. Granite is a typical example of this type of rock.

acquired immune deficiency syndrome see **AIDS**.

acre a measure of area, equal to 4840 square yards (4047 square metres or 0.4 hectares).

acrylic resins a class of THERMOPLASTICS formed by polymerization of AMIDES or ESTERS. They are colourless and transparent and are resistant to most weak chemicals, ageing and light. They are commonly used for optical products, e.g. lenses. Trade names include Perspex.

actinides the name for a group of elements in the PERIODIC TABLE with ATOMIC NUMBERS from 89 (actinium) to 103 (lawrencium) inclusive. All the elements resemble actinium and are radioactive and many do not occur naturally.

action potential a biochemical change produced in a nerve by some sort of stimulus, such as pain. The stimulus is changed into a signal that moves along the nerve and this is achieved by the movement of sodium ions across the membrane of the nerve cells. The movement of the sodium ions, which are positively charged (Na^+) in effect creates an electrical pulse along the nerve.

activation energy a chemical reaction between compounds has to be started in some way and the energy required is called the activation energy. In some reactions, the KINETIC ENERGY of the molecules reacting together, that is the energy due just to the movement of the molecules, is enough to start the reaction. However, in many cases the activation energy is too high for the reaction to start by this means and others methods have to be employed. This may involve putting energy into the system by way of heat, or a CATALYST may be used.

acute in trigonometry, the term used to describe an angle that is between 0° and 90°.

adaptation is basically the change which occurs in an organism or animal in response to its environment and the conditions in which it lives. This process of change can take a very long time but results in the animal being suited to a particular diet, developing a certain way of finding food, living in hostile conditions and so on. The change can also be genetic, as when the organism possesses GENES that produce characteristics that prove to be beneficial for survival in its environment. A striking example of genetic adaptation is the change that occurs in populations of bacteria which have become resistant to penicillin due to the widespread use of this drug in the past. Such antibiotic resistance is a major problem in hospitals when trying to prevent any risk of infection after an operation.

Adaptation can be seen in many other ways, especially the variety in birds' bills (their beaks) which enables them to obtain their particular food. Plants also show a tremendous range of adaptations, an obvious example being the cactus which has adapted so that it can survive in the harsh heat of the desert.

adaptive radiation is a response to adaptation in which a species of animal develops through evolution into numerous descendent species which can use to the full the various habitats that exist throughout the world. This evolutionary spreading out of species probably explains the bewildering array of, say, amphibians, which has arisen as a result of adaptive radiation after the first amphibians moved on to land.

addition formula an equation used to find the sum or difference of angles as a sum or difference of the products of the trigonometric functions of the individual angle. For example, the formulae for the functions of cosine and sine are

cosine $(A+B) = \cos A \cos B + \sin A \sin B$

sine $(A+B) = \sin A \cos B + \cos A \sin B$

addition polymerization the formation of a large molecule called a polymer, in which the structure consists of the same atoms joined together and repeated in a long chain. The reaction involves the addition of the basic building block molecule (called the monomer) to form the chain. Addition polymerizations often occur at high temperatures and pressure in the presence of a CATALYST.

adenosine triphosphate SEE **ATP**.

adiabatic change a change that occurs with no alteration in the heat content of the system.

ADP see **ATP**.

adrenal gland see **endocrine system**.

adsorption the taking up, or concentration of, one substance at the surface of another, e.g. a dissolved substance on the surface of a solid.

aerobic respiration a set of reactions that occur in plant and animals cells in which food molecules are broken down to form energy. The key feature is that oxygen is required for this process. The reactions are catalysed by ENZYMES.

aerosol a fine mist or fog which is dispersed in and carried by a gas. The particles may be liquid or solid. The term also describes the pressurized container with a spray mechanism which is used extensively for deodorants, insecticides, etc. The typical can of aerosol consists of the liquid to be sprayed mixed with a gas that forces the liquid out when the button is pressed and a valve is opened.

AIDS (*short for* Acquired Immune Deficiency Syndrome) a serious disease thought to be caused by a human retrovirus called HIV, human immune deficiency virus. This seriously affects and impairs the immune system and leaves the patient open to both minor and major infections, as well as the possibility of developing cancer. Not all people infected with HIV develop AIDS, but it can be passed on by the following methods: the receiving of infected blood during transfusions, although blood is now treated to avoid this possibility; an infected mother may pass it on to her child during pregnancy; the sharing of needles in drug abuse; and the passing of body fluids during sexual contact. There is no evidence that HIV can be transmitted by everyday activities and social contact, such as swimming, sharing cutlery, using a public toilet, etc.

alcohol one of a group of organic compounds that are made up of carbon, oxygen and hydrogen. They are similar in structure to the ALKANES but have an oxygen/hydrogen group, -OH (hydroxyl), attached to the molecule, e.g., methanol is CH_3OH. Alcohols all occur in liquid form and have many uses. Alcoholic drinks contain ETHANOL (C_2H_5OH), the alcohol obtained during the fermentation of sugars or starches and many alcohols are used as SOLVENTS.

aldehyde (*also known as* **alkanal**) one of a group of organic compounds made up of carbon, hydrogen and oxygen which contain a carbon/oxygen group (called a CO radical) attached to both a hydrogen atom and a hydrocarbon group, giving a formula of the type R.CO.H, so if R is H, we get HCHO which is the formula for formaldehyde, more correctly known as methanal.

alga (*plural* **algae**) the common name for a simple water plant, which is without root, stem or leaves. Algae range in form from single cells to plants many metres in length. The blue-green algae, cyanobacteria, are widely distributed in many environments. Algae are very important ecologically being responsible for a significant proportion of the products of photosynthesis and in both fresh and marine water they provide a food source for many organisms. They contain chlorophyll as found in other plants but in addition contain different forms of chlorophyll which, through their various colours, provide a ready way of grouping algae.

algebra the use of symbols, particularly letters, to help solve mathematical problems where the object is to find the value of an unknown quantity or to study complex systems and theories. In its simplest form, it may involve finding an unknown quantity in an equation such as: $5x - 3 = 22$, where x is clearly 5. Albert EINSTEIN used advanced algebra to derive equations from his general theory of RELATIVITY.

aliphatic the descriptive term for organic compounds that contain carbon atoms in open chains (unlike AROMATIC HYDROCARBONS). In addition to the main groups e.g. alkanes, the term includes all products derived from such compounds and substitution products (i.e., compounds made by the replacement of an atom or group of atoms in the aliphatic molecule).

alkali a chemical (see also BASE) which can be thought of as the opposite to an acid because in the form of a solution it will give a pH value greater than 7. The hydroxides of the metallic elements sodium (Na) and potassium (K) are strong alkalis, as is ammonia solution (ammonium hydroxide, NH_4OH).

alkali metals the metals lithium (Li), sodium (Na), potassium (K), rubidium (Rb), and caesium (Cs), which belong to group IA of the PERIODIC TABLE. The last member of the group, francium (Fr), occurs only as a radioactive isotope. All the metals have a valency of one, that is they have one electron in their outer shell (or electron orbit) which is commonly used in creating chemical bonds with other elements. The elements are all prepared from their HALIDES (a compound with chlorine, bromine etc., occurring as a chloride, bromide respectively) by a process called ELECTROLYSIS. Their melting and boiling points fall with increasing atomic weight, so that caesium has the second lowest melting point of any metal.

alkaloids organic substances found in plants which give the plant some protection against being eaten. Many alkaloids are also used in medicine, for example codeine and morphine.

alkanal see **aldehyde**.

alkane a compound containing hydrogen and carbon in an open chain structure with single bonds between each carbon atom, to which the hydrogens are attached. A series of such compounds occurs and the first member is METHANE, CH_4. Further members can be thought of as the basic member, methane plus the unit $-CH_2$. The general formula for the series is C_nH_{2n+2}, thus the chemical formula for the second member of the series, ETHANE, is C_2H_6, and after that comes C_3H_8, C_4H_{10}, etc. Alkanes are called saturated compounds and because of this they are quite stable.

alkene compounds similar to the ALKANES in that they consist of hydrogen and carbon in an open chain structure but with double bonds between the carbon atoms. The first member of this series is ETHENE, C_2H_4, and all other members follow the general formula C_nH_{2n}. Alkenes are called UNSATURATED hydrocarbons, because they can easily undergo reactions which allows the double bond to be replaced by a single bond and other atoms to be joined onto the molecule.

alkyne with alkanes and alkenes, a third group of compounds containing hydrogen and carbon in an open chain structure. The carbon atoms are joined together by triple bonds and the first member of the alkynes is ETHYNE (also called acetylene), C_2H_2, and the general formula for the other members is C_nH_{2n-2}. They are UNSATURATED compounds that will readily undergo reactions across their triple bond.

allergy (*plural* **allergies**) results from an excessive response of the immune system of the body to foreign substances. The body responds to harmless substances, such as pollen, but because the body is oversensitive to such substances (which are called antigens or allergens) it releases compounds called histamines in an effort to destroy the substance.

This causes the characteristic symptoms of an allergy, i.e. inflammation, itching etc. In addition to pollen, an induced overreaction can be produced by certain foods, or even toxins injected by insects such as wasps. Such reactions can be counter-attacked through the use of antihistamine drugs. Hay fever is probably one of the commonest allergies and it is caused by the pollen from grasses and plants.

alpha decay during radioactive decay, the emission of ALPHA PARTICLES from the nucleus of a radioactive ISOTOPE.

Alpha Centauri the brightest of three stars in the constellation Centaurus. Another of the three, Proxima Centauri, is the nearest star to our SUN.

alphanumeric a set of characters derived from the numerals 0 to 9 and the alphabet. In computing, the remaining keyboard characters are used for functions other than keying text.

alpha particle a fast-moving helium nucleus ($^4_2He^{2+}$) with a short, straight range from the source of emission.

alternating current (AC) an electric current that reverses the direction of its flow regularly, resulting in a constant frequency independent of the type of CIRCUIT. Mains electricity is made up of alternating current as opposed to DIRECT CURRENT and in effect the current is pushed and pulled through the circuit 50 times per second (50 Hz) and in the UK the alternating current is supplied for domestic users at a voltage of 240v.

altitude in geometry, the distance measured by a line from the corner (vertex) of a figure to the opposite side of that figure, e.g. the height of a triangle; from one corner to the opposite line (base).

altocumulus grey-white sheets, banded layers or rolls of cloud in small segments which occur at heights of 3000 to 7500 metres.

altostratus grey sheets or layers of cloud which may have a fibrous or uniform appearance. They occur at heights of 3000 to 7500 metres and may be thin enough to allow sunshine through.

aluminium a light metal that is easily shaped and drawn out into wire and which is also a good conductor of electricity. It is the most common metallic element in the earth's crust (third most common element overall, at 8 per cent). It is extracted from its ore (because it never occurs as pure aluminium) bauxite by electrolysis with molten cryolite as a FLUX. When left in air, the surface reacts to form a layer of oxide (alumina) which prevents corrosion. The metal and its alloys are used for many purposes, including the manufacture of cooking utensils, electrical equipment, aircraft, foil, cans for drinks.

amalgam the alloy, or mixture, of a metal with another metal, mercury. Most metals will mix with mercury which in its natural state is liquid. One of the commonest amalgams (silver/mercury) was used widely in dentistry for fillings, but this is being replaced by modern materials.

American Standard Code for Information Interchange see **ASCII**.

amines compounds formed from ammonia, NH_3, by the replacement of one or more hydrogen atoms to give three classes: primary, NH_2R; secondary, NHR_2; and tertiary, NR_3. In these cases the R is an organic group such as the hydrocarbon CH_3, which forms methylamine, $CH_3 NH_2$.

amino acid any of the 20 standard organic compounds that serve as the building blocks from which all proteins are created. All have the same basic structure, containing an acidic carboxyl group (-COOH) and an amino group (-NH₂), both bonded to the same central carbon atom referred to as the α-carbon. Their different chemical and physical properties result from one group which changes, the side chain or R-group, which is also attached to the α-carbon. The simplest amino acid is glycine,

where the lower hydrogen is the R-group. In a diagram of a protein structure glycine is abbreviated as gly. The remaining nineteen are:

alanine (ala)	cysteine (cys)
valine (val)	tyrosine (tyr)
leucine (leu)	asparagine (asn)
isoleucine (ile)	glutamine (gln)
methionine (met)	aspartic acid (asp)
phenylalanine (phe)	gluamic acid (glu)
tryptophan (trp)	lysine (lys)
proline (pro)	arginine (arg)
serine (ser)	histidine (his)
threonine (thr)	

amino group an essential part of the AMINO ACIDS. It consists of a nitrogen atom bonded to two hydrogens (NH_2) which is then bonded to the central carbon of the amino acid.

ammonia a compound (written as NH_3) that exists as a colourless gas. It will react with water, giving the alkaline solution known as ammonium hydroxide (NH_4OH). Ammonia will ionize in water to form the ammonium ion (NH_4^+) and the hydroxide ion (OH-)

$$NH_3 (g) + H_2O (l) \rightarrow NH_4^+(aq) + OH^-(aq).$$

Ammonia has an irritating smell and is used a great deal in industry. Liquid ammonia is used for refrigeration and as a solution, ammonia is a cleaning agent. Many industrial chemicals and compounds are manufactured from ammonia, including fertilizers and explosives.

amorphous having no shape or form. The term used for a substance when it is non-crystalline.

ampere (A) the unit for measuring the quantity of electric CURRENT, often abbreviated to **AMPS**. It shows the rate at which charge is flowing around the circuit.

amphiboles a major group of rock-forming minerals that are silicates with iron, calcium, magnesium, sodium and aluminium. They occur in many different rock types, particularly igneous and metamorphic rocks and hornblende is one of the commonest amphiboles.

amphoterism a property of the oxides or hydroxides of certain metallic elements, which allows them to function both as acids and alkalis. Although insoluble in water, amphoteric compounds will dissolve in acidic solutions (pH < 7) or basic solutions (pH > 7) to form a salt. An example is aluminium hydroxide, Al $(OH)_3$.

amplitude the maximum displacement of a particle from its position of rest.

For sound waves, the amplitude relates to the intensity of the wave and is the distance between the x-axis (rest position) and the crest or trough of the wave. In a simple pendulum, the amplitude is defined as the angle through which the arm moves when swinging between its extreme and rest positions.

AMPS see **ampere**.

anaerobic respiration one form of respiration in which biochemical reactions occur to release energy, but the reactions occur without oxygen being present. The energy released is less than that of AEROBIC respiration. It occurs in just a few groups of organisms, such as bacteria in stagnant ponds. It also happens in muscles when oxygen is absent, to produce lactic acids, resulting in the familiar feeling of cramp.

anemometer a device which measures the speed of the wind and often consists of 4 cups at the end of arms—the rotating cups anemometer. There is also a pressure driven variety which relies upon the wind pressure through a tube (pressure-tube anemometer).

angström the unit of measurement (10^{-10}m) which has now been replaced by the nanometre ($10\text{Å} = 1\text{mm}$), and which was used for electromagnetic radiation. It was also used by crystallographers for measurements between atoms in crystal structures.

angular momentum the momentum of a body rotating around a point. The angular momentum of a body is the product of the angular velocity (its motion through an angle about an axis) and the moment of inertia (the mass of the body multiplied by the square of its distance from the axis). The Earth has both rotational angular momentum (because it rotates on its axis) and orbital angular momentum (because it orbits the sun).

anion a negatively charged ion formed by an atom or group of atoms gaining one or more electrons. For example a chloride ion is Cl. Anions may occur in the solid state or in solutions or melts (e.g molten salts).

anisotropy the term used to describe a substance when it possesses a property which is related to a particular direction within the substance, e.g., some crystals have a different REFRACTIVE INDEX in different directions. Anisotropy is the opposite of ISOTROPY.

annealing the process whereby materials, usually metals, are heated to, and held at, a specific temperature before controlled cooling. This relieves STRAIN set up by other processes.

annulus the shape created by the area between two concentric circles, as with a washer.

anode in an electrochemical cell, the anode is the positively charged electrode to which the anions of the solution move to give up their extra electrons,

anodizing the process in which a coat of oxide is deposited on the surface of a metal (often aluminium or an alloy) by making the metal the ANODE IN ELECTROLYSIS. The oxide skin forms a protective layer and may be made decorative through the use of a dye in the electrolytic process.

antibiotic a chemical produced by micro-organisms, such as BACTERIA and moulds, that can kill bacteria or prevent their growth. The first antibiotic to be discovered was penicillin, and there are now many more, including erythromycin, streptomycin and terramycin. They are used extensively to treat infections but using them too much can weaken the natural defences of the body and it is possible for the micro-organisms to develop into strains that can resist the antibiotics.

antibody (*plural* **antibodies**) a protein circulating in the blood, which is produced by special white blood cells when a foreign substance (antigen) enters the body. The production of antibodies is called an *immune response* and antibodies consist of protein chains that form IMMUNOGLOBIN (shortened to Ig), and although millions of different antibodies are produced in order to cope with any micro-organism that causes disease, there are only five major classes, which have the following functions:

Antibody	Function
IgG	The most abundant, it fights micro-organisms and toxins. It is the first immuno-globin found in newly born infants.
IgA	The major Ig that defends surfaces in contact with the outside, including the gut wall.
IgM	The first Ig to be produced during infection, it is very effective against bacterial infections.
IgD	No specific function is known for this immunoglobin.
IgE	This protects external surfaces and triggers the release of HISTAMINE from other cells of the immune system.

anticyclone an area of pressure which increases to and reaches its highest at the centre. Winds tend to be light, because there are small pressure gradients and flow clockwise in the northern and anticlockwise in the southern hemisphere. Associated weather tends to be settled and fine although anticyclones in winter can give rise to cold conditions and often FOG.

antigen any substance that the body sees as being foreign. This triggers an immune response from the body's IMMUNE SYSTEM. Common antigens are PROTEINS present on the surface of bacteria and viruses. Unsuccessful transplant operations are usually a result of the patient's immune response recognizing the surface cells of the organ from the donor as non-self. The organ is said to be rejected when the patient's immune system becomes activated and tries to destroy the donated organ.

antioxidant a substance added to some materials, such as paint, oils and rubber, to delay the harmful process of oxidation. They are also added to foods, particularly fats.

aorta the largest ARTERY in the body and through it the blood flows out of the left VENTRICLE of the heart. The aorta is approximately three centimetres across, with thick muscular walls to carry blood under pressure. The aorta divides into several branches, which supply blood to the arms and the head. It then continues down around the spine to the level of the lower abdomen, where it divides into two major branches to supply the legs.

apatite a mineral that contains calcium phosphate with fluorine, chlorine, and HYDROXYL (oxygen and hydrogen, OH) ions. It occurs in small amounts as a mineral in IGNEOUS and METAMORPHIC rocks and is the main constituent of fossil bones and also the enamel of vertebrate teeth. It is used industrially in the manufacture of fertilizers.

apogee the point at which a satellite is at its greatest distance from the earth in its orbit around the Earth. The opposite situation is called the perigee.

aquifer a rock containing pores into which water can move, such that when this rock is underlain by impermeable strata, it may contain significant quantities of water which can be extracted.

Archimedes' principle a law of physics stating that when a body is partly or totally immersed in a liquid, the weight it appears to have lost is equal to the weight of the displaced liquid.

argon one of a group of gases called the INERT gases. It is unreactive and is used in lamps, fluorescent tubes, and as an unreactive, inert shield in arc welding.

arithmetic mean an average. For a set of numbers with *n* values the mean is the sum of the numbers, divided by *n*. For example, the mean of six numbers 2, 4, 3, 9, 5 and 1 is their sum, 24, divided by the number of values which is six, giving four.

arithmetic series the sum of the terms in a sequence of quantities (an arithmetic sequence). An arithmetic sequence could be 2, 5, 8, 11 and 3 would be the COMMON DIFFERENCE. An unknown term (called the nth term) can be found from:

nth term = $a + (n - 1)d$

where a is the first term. So if the first term is 2 and the common difference, d, is 3, the eighth term would be $2 + (8-1)3 = 2 + 21 = 23$.

The sum of n terms is calculated using:

$$S_n = n/2 \, [2a + (n - 1)d].$$

armature the rotating wire coil of an electric motor. More generally, it is any electric component within a piece of equipment in which a VOLTAGE is induced by a MAGNETIC field.

arteriosclerosis the thickening, hardening and loss of elasticity of the ARTERIES. This can be a condition of advancing age, or it can be associated with fatty deposits, particularly CHOLESTEROL, that block the arteries, causing their diameter to decrease. As a result, the heart must strain to increase its muscular activity to generate enough pressure to pump the blood through the arteries.

artery (*plural* **arteries**) a thick-walled vessel that carries blood under pressure resulting from the pumping mechanism of the heart. Arteries carry oxygenated blood from the heart to body tissues. The one exception is the PULMONARY ARTERY which carries deoxygenated blood from the heart to the lungs.

artificial intelligence the concept that computers can be developed to work in a way that is similar to human intelligence including learning, reasoning and adaptation. Also, a branch of computer science including robotics.

ASCII (*acronym for* **A**merican **S**tandard **C**ode for **I**nformation **I**nterchange) a standard code of 128 ALPHANUMERIC characters for storing and exchanging information between computer programs.

ascorbic acid another name for vitamin C which is found in all citrus fruits and green vegetables (especially peppers). A lack of ascorbic acid leads to the fragility of tendons, blood vessels and skin, all of which are characteristic of the disease called scurvy. The presence of ascorbic acid is also believed to help in the uptake of iron during the process of digestion by the body.

assay the analysis of a mixture by chemical means to determine the amount of a particular constituent, e.g. the amount of metal in an ore.

asteroid one of many rocky or metallic bodies, more correctly called planetoids, that orbit the sun between the orbits of Mars and Jupiter in the asteroid belt. Most are very small, down to the size of dust, but the largest, Ceres, has a diameter of about 1,000 kilometres. Asteroids are believed to be remnants from the formation of the solar system. Meteorites are debris from the asteroid belt formed by the collision of the bodies.

astronomical unit (AU) the average distance from the centre of the EARTH to the centre of the SUN and equal to 1.496×10^8km (92.9×10^6 miles). The AU is used as a measure of distance within the solar system.

atmosphere a unit of PRESSURE defined as the pressure that will support a mercury column 760mm high at 0°C, sea level, and a latitude of 45°. Also, the layer of gases surrounding the earth, which contains, on average, 78 per cent nitrogen, 21 per cent oxygen, almost 1 per cent argon, and then very small quantities of carbon dioxide, neon, helium, krypton and xenon. In addition, air usually contains water vapour, hydrocarbons and traces of other materials and compounds. At higher altitudes, the atmosphere thins due to a reduction in pressure and the temperature falls.

atom the smallest particle that makes up all matter and yet still retains the chemical properties of the element. Atoms consist of a minute nucleus containing PROTONS (p) and NEUTRONS (n), with negatively charged particles called ELECTRONS (e) moving around the nucleus in orbits which spread out into clouds in which the electrons move at very high speeds. The number of protons in an atom is equal to the number of electrons, and as the protons are positively charged, the atom is, overall, electrically neutral. The various elements of the PERIODIC TABLE all have a unique number of protons within their atomic nucleus.

atom bomb see **nuclear fission**.

atomic mass unit (see RELATIVE ATOMIC MASS) defined in 1961 as one twelfth of the mass of an atom of ^{12}C, having formerly been based upon ^{16}O, the most abundant isotope of oxygen.

atomic number (A, at. no.) the number of protons in the NUCLEUS of an ATOM. Although all atoms of the same element will have the same number of protons, they can differ in their number of neutrons, resulting in an isotope.

atomic weight see **relative atomic mass**.

ATP (*abbreviation for* adenosine triphosphate) an important molecule that is used as an energy source to drive all processes that occur in biological cells. ATP is used continuously when an organism performs work or undertakes biochemical processes, and an enormous amount of ATP is required by, for example, a working muscle. However, it can be regenerated within cells at an astonishing rate.

atrium (*plural* atria) a minor chamber of the heart that is considered to be a reservoir as blood passes from it into the pumping chamber, the VENTRICLE. The right atrium of the heart receives the blood carried by the superior and inferior VENA CAVA before it passes via a valve into the right ventricle.

The pulmonary veins carry oxygenated blood from the lungs into the left atrium, which then flows via a valve into the left ventricle.

aureole in geology, the *contact* or *metamorphic* aureole is the area around an igneous intrusion where the host rocks have been subjected to the heat of the intrusion. This results in a melting or partial melting of the host rocks and a recrystallization often with the growth of new minerals.

In meteorology, a ring (the inner part of the CORONA) sometimes seen around the Sun or Moon. The effect is created by DIFFRACTION of light by water droplets in high cloud formations.

aurora luminous and often colourful sheets or streaks in the sky, formed by high-speed, electrically charged solar particles entering the upper atmosphere and bombarding atoms and molecules. Electrons are then released causing an associated release of energy as light. These effects are related to SUNSPOT activity and are termed the Northern Lights (*aurora borealis*) in the northern hemisphere, and the Southern Lights (*aurora australis*) in the southern hemisphere.

Avogadro's constant or **Avogadro's number** the number of particles present in one mole of a substance. It is given the symbol N or L and has the value of 6.023×10^{23}.

Avogadro's law the principle formulated by the Italian scientist Amedeo Avogadro (1776-1856) which states that equal volumes of all gases contain the same number of molecules when under the same temperature and pressure. For the purpose of calculation, one MOLE of gas will occupy a volume of 22.4 litres at standard conditions, i.e. 273.15K and 1 atmosphere.

B

bacillus (*plural* **bacilli**) a BACTERIUM which has a rod-shaped form.

background 'interference' which affects the reading of a signal that is being measured. The interference signals come from sources such as natural RADIOACTIVITY and COSMIC RAYS and must be allowed for when measurements are being taken.

backscatter when radiation strikes a surface, some of the beam is REFLECTED back again towards the source. This is called backscatter.

bacteriophage or **phage** a virus that attacks and infects a BACTERIUM.

bacterium (*plural* **bacteria**) a MICRO-ORGANISM (one that cannot be seen with the naked eye but only with the aid of a MICROSCOPE), which usually has a body made of only one cell (unicellular). Bacteria are believed by many scientists to be the first organisms to have existed on Earth and they have always been extremely important in all life processes. Bacteria occur in water, air, soil and rotting plant or animal debris, and even in extremely inhospitable environments such as hot springs full of chemicals. They are vitally important in the breakdown and decomposition of ORGANIC material—without bacteria, nothing would rot away. Most bacteria need oxygen in order to live and are called *aerobic*, but others do not and are termed *anaerobic*. The surrounding layer around a bacterial cell is called the cell wall. Two types of bacteria, which each have a different cell wall structure (Gram positive or Gram negative) are identified by a test known as Gram's stain. Some bacteria have a protective, slimy outer layer called a capsule, and others may have hairs called filaments which cause movement. Bacteria occur in a number of different shapes and forms which help to identify them. These are spiral (*plural* spirilli, *singular*, spirillus), spherical or round (*plural* cocci, *singular* coccus), rod-like (*plural* bacilli, *singular*, bacillus), comma-shaped (vibrio) and corkscrew-shaped (spirochaetae). Bacteria usually reproduce by asexual reproduction and a few are responsible for extremely serious diseases in plants, animals and man, e.g. typhoid, tuberculosis, diphtheria and cholera. These diseases can be treated with ANTIBIOTICS although there is a problem of resistance of some bacteria to certain antibiotic drugs.

ballistics the study of the flight path of an object which is being influenced by GRAVITY as it moves. It is applied to bullets, rockets, missiles etc.

bar chart or **bar graph** a graph that illustrates the relationships between two variable properties by vertical parallel bars. The height of the bars is proportional to the variation in the data.

barometer an instrument which measures the pressure that the atmosphere exerts on the Earth's surface. Changes in atmospheric pressure indicate that alterations in the weather are likely to occur.

barrier reef a reef of coral built up parallel to the shore but some distance from it so that a lagoon is created between the two. An example is the Great Barrier Reef, Australia, which is almost 2000 kilometres (1243 miles) long.

basalt a dark, fine-grained type of igneous rock associated with volcanic activity. Lava flows of basalt cover over two thirds of the Earth's surface, both on land and beneath the sea.

base in a chemical reaction, any substance that dissociates (breaks up) in water to produce hydroxide (OH) IONS.

In mathematics, a base is the number raised to a certain power (EXPONENT), which will produced a fixed number. For example, base 5 to the power of 3 equals 125, i.e. $5^3 = 5 \times 5 \times 5 = 125$.

Beaufort scale a scale indicating wind velocities determined from measurements taken at a height of 10 metres (32.8 feet) above ground level. This numerical scale ranges from 0 (calm, speed 0.3 ms-1) to 12 (hurricane, speed 32.7 ms^{-1}).

Becquerel the unit of radioactivity in the SI scheme which is named after Antoine Henri Becquerel (1852–1908), who began the study of radioactivity.

bedding plane in geology, the surface that separates each bed in a sequence of SEDIMENTARY ROCKS. It represents a break or change in conditions during the deposition (or laying down) of the particles or grains from which the rocks were eventually formed.

benthic a term which describes any plant or animal living on the bottom of a lake or sea.

benzene an organic chemical compound, which is a HYDROCARBON containing carbon and hydrogen, having the formula C_6H_6. It has a ring structure formed from its six carbon ATOMS and at room temperature is a colourless liquid with a distinctive smell. It is a toxic (poisonous) and carcinogenic (may cause cancer) substance which is very important in the particular branch of chemistry called organic chemistry.

beri-beri a serious disease of human beings caused by a lack or deficiency in the diet of vitamin B (thiamine).

beryl a mineral which is found in some IGNEOUS and METAMORPHIC rocks which, when coloured, is regarded as a precious stone. Chemically it is beryllium alumino-silicate with the formula $Bl_3Al_2Si_6O_{18}$.

bicarbonates chemical substances which are acid salts of carbonic acid (H_2CO_3) in which one hydrogen atom is replaced by a metal.

bilateral symmetry describes any organism which, if divided lengthwise down the middle, would be split into two halves that are almost mirror images of each other.

bile a thick fluid made in the liver and stored in a small pouch, called the gall bladder, near the liver. It aids the digestion of fats by breaking down large fatty particles into smaller ones in a part of the small intestine called the DUODENUM. When bile is needed, the gall bladder contracts (becomes smaller) by means of muscles in its wall, and forces the fluid into the duodenum through a pipe called the bile duct.

binary system a type of code in arithmetic which uses a combination of the two digits 0 and 1, expressed to the base 2. Therefore, beginning with the value 1 in the right-hand column, each move to the left sees an increase to the power of 2.

1×2^5	1×2^4	1×3^3	1×2^2	1×2^1	1×2^0	origin of binary numbers
32	16	8	4	2	1	binary numbers
1	0	1	1	1	0	example
Hence in the example, the figure is $32 + 8 + 4 + 2 = 46$						

The binary number system is the basis of code information in computers, as the digits 0 and 1 are used to represent the two states of off or on in an electronic switch in a circuit.

biochemistry the scientific study of the chemical processes which take place within the CELLS and TISSUES of all living things.

biosphere the part of the Earth's surface, both land and water and including part of the atmosphere, which is inhabited or could be inhabited by any living organism.

biosynthesis the production, by the CELLS of living organisms, of complicated chemical substances. ENZYMES are necessary for these processes to take place.

biotechnology the harnessing by man of the ability of organisms to produce drugs, food or other useful products. MICRO-ORGANISMS are the main ones involved in biotechnology, especially BACTERIA and FUNGI. Examples include the fermentation of yeast to produce alcohol and in the making of bread, and the growth of certain fungi to make antibiotic drugs. More recently, GENETIC ENGINEERING or the altering of the GENES, the building blocks which determine the make-up of an organism, has increasingly been used in biotechnology.

biotic anything relating to life or living things.

bit an abbreviation for binary digit, either the number 1 or 0. (See BINARY SYSTEM).

black hole a region in space from which no material or light escapes due to its enormous gravitational force. Black holes are thought to be the remains of massive stars which have come to the end of their life and collapsed in upon themselves.

blood a vital substance consisting of red blood cells (ERYTHROCYTES) and white blood cells (leucocytes) suspended in a liquid called BLOOD PLASMA. The blood of mammals contains many PROTEINS which are involved in blood clotting. Blood circulates around the body in ARTERIES and VEINS and acts as a transport system for many substances. These include oxygen, the end products of digestion including amino acids (proteins), lipids (fats), sugars, glucose (carbohydrates), hormones, and waste products (AMMONIA and CARBON DIOXIDE).

blood clotting (also called haemostasis) a process, involving several chemical substances and reactions that stops blood from leaking out of an area of injured tissue. At first the blood vessels in the injured parts contract and become narrower so that less blood escapes. Then a plug or clot is formed to further seal off the damaged part, peventing micro-organisms which might cause infection from entering. The plug or clot is formed in response to the release of an enzyme by the damaged blood vessels and blood platelets. This enzyme triggers off a series of changes resulting in the production of a firm clot which contains platelets, trapped red blood cells and a PROTEIN called fibrin.

blood grouping a method for classifying blood types by checking which particular antigens (proteins) are present on the surface of red blood cells. There are many systems for classifying blood types but the two most important use the ABO and rhesus blood groups. The ABO blood group system is based on the presence or absence of two antigens called A and B. The A and B antigens may be present alone, to give blood groups A or B, together, to give AB, or both may be absent, resulting in blood group O. Hence there are four blood groups, A, B, AB or O.

The rhesus blood group system can be simply explained in terms of whether a person has the rhesus factor (Rh) or D antigen present on the surface of his or her red blood cells. If the Rh factor is present the person is Rhesus positive and if is not, he or she is rhesus negative.

blood plasma blood from which all the blood cells (red cells, white cells and platelets) have been removed. The liquid which remains, i.e. the plasma, is 90% water and contains some PROTEINS, sugar, salt, UREA, HORMONES and VITAMINS.

blood serum plasma from which one of the PROTEINS, called FIBRINOGEN, has been removed.

boiling point the temperature at which a substance changes from the liquid state to the gaseous state.

bond the force that holds ATOMS together to form a MOLECULE.

bond dissociation energy the amount of energy needed to break a particular bond holding two atoms together.

bone a hard material which forms the skeleton of most vertebrate animals. It is a type of connective tissue consisting of collagen fibres (a type of protein), bone salts and bone cells.

botulism the most dangerous type of food poisoning in the world caused by a BACTERIUM called *Clostridium botulism*. This is an anaerobic type of bacterium, i.e. one which does not require oxygen to live. It can exist in such conditions as an airtight food can and, during growth, releases poisons called toxins which may be fatal to human beings.

boulder clay a type of deposit of rocks embedded in clay left by a GLACIER or ice sheet which has since melted or retreated. It contains rocks picked up by the ice as it travelled over an often extensive area. In Britain, boulder clays occur which were left when the ice retreated following the last Ice Age or GLACIATION.

Boyle's law a law of physics devised by the Irish scientist Robert Boyle (1627–1691) which states that at a constant temperature, the volume of a gas lessens in proportion to an increase in the pressure of the gas. Hence at a constant temperature, if the pressure is doubled the volume is halved. (*See also* GAS LAWS).

brackish a term that is used to describe water that is half or part way in saltiness between the sea and freshwater, as occurs in estuaries. (*See* ESTUARY).

brain a major collection of nerve TISSUE in vertebrate animals which, along with the SPINAL CORD, forms the central nervous system. It receives and decodes information which comes to it from both outside and inside the animal, and sends out information to all parts of the body such as the muscles. Many INVERTEBRATE animals also possess a simple type of brain.

bronchus (*plural* **bronchi**) one of two tubes which divide off from the windpipe or trachea in vertebrate animals. These divide into further smaller tubes which end in the LUNGS.

Brownian motion an event first discovered by a Scottish botanist, Robert Brown (1773–1858) in 1827. He observed that a random movement of minute particles occurs in both gases and liquids. (*See* KINETIC ENERGY).

budding a form of asexual reproduction in animals in which part of the parent develops a bulge (bud) which becomes detached to form a new organism. Budding only occurs in primitive organisms which have a fairly simple structure, such as sponges.

In plants, budding is found in yeasts and other single-celled fungi. In horticulture, bud grafting of woody plants is often carried out, especially in fruit trees and roses. A bud or shoot called a *scion*, together with some of the stem beneath, is placed inside a cut or *graft* through the bark of a host plant, the *stock*. The stock may be a closely related type of plant to the scion, or a wild species. The two become one plant once the procedure has been carried out.

buffer a chemical substance which is able to keep the pH (the amount of acidity) of a solution (liquid) at more or less the same level when other substances are added. It does this by either mopping up or releasing hydrogen ions (H+) as needed. It is the level of hydrogen ions which determines the pH.

Bunsen burner a gas burner invented by the German scientist Robert Wilhelm Bunsen (1811–1899), and widely used in science laboratories, especially in chemistry. The small upright tube has an adjustable air inlet at the base which allows the size of the flame to be controlled. The flame is produced by burning a mixture of HYDROCARBON gas and air, and has an inner cone where CARBON MONOXIDE is formed and an outer fringe where it is burnt.

burette a piece of equipment used in chemistry consisting of a narrow glass tube, fixed upright with a tap at the bottom. The tube is graduated (marked with a scale) so that measured volumes of liquid can be released by means of the tap. (See TITRATION and VOLUMETRIC ANALYSIS).

butane a chemical substance with the formula C_4H_{10} which is very useful as it is easily changed from a gas to a liquid, allowing it to be stored and used as a fuel.

byte in computing, usually a sequence of 8 or 16 bits representing one character or a unit of memory. The memory capacity of a computer or the amount of its memory, is measured in thousands of bytes (called kilobytes, KB) or millions of bytes (called megabytes, MB).

C

caffeine a naturally-occurring chemical substance that is found in tea leaves, coffee beans and other plants. It acts as a weak stimulant on the central nervous system so after drinking a cup of coffee or tea, a person may feel more alert.

calcium a metallic substance with the symbol Ca which is common in nature being found in both plants and animals. It is essential for the normal growth and development of animals, being found in bones, teeth, blood and nerves.

calcium carbonate a common chemical compound in nature with the formula $CaCo_3$. In the form of calcite, it is an abundant MINERAL in rocks especially LIMESTONE, CHALK and MARBLE. It is used by man in the manufacture of cement and fertilizer.

calculus a large branch of mathematics concerned with quantities which are continuously variable. The many techniques of calculus were developed by the British scientist Isaac Newton (1643– 1727) and the German philosopher Gottfried Wilhelm Leibniz (1646–1716).

calorie a unit of quantity of heat defined as that heat needed to raise the temperature of one gram of water through 1°C. Nowadays, it has largely been replaced by the JOULE (1 calorie = 4.186 joules).

calorimetry the discovery of certain thermal properties of substances such as calorific value, specific heat or latent heat in physics and chemistry. The instrument used is called a calorimeter and it consists essentially of apparatus which allows the substance to be burnt and the heat transferred to a surrounding body of water, enabling the rise in temperature and therefore the heat output to be measured.

Calvin cycle a series of chemical reactions that take place in some plant cells during the last stage of PHOTOSYNTHESIS, named after the American biochemist Melvin Calvin.

cancer a disease in which the characteristic feature is the uncontrolled and rapid growth of CELLS, leading to the formation of lumps of abnormal tissue, called tumours. Some tumours are called *benign* because they are usually harmless (although they may still need medical treatment). Others are called *malignant* because cancer cells break off from the tumour and travel in the bloodstream to grow again somewhere else in the body. These disrupt the normal function of the cells and tissues where they grow and may eventually cause death. However, many types of malignant cancer can be effectively treated by modern medical methods, which include surgery, treatment with drugs (chemotherapy) and radiation (radiotherapy). A great deal of scientific research is carried out into the causes of cancer and, in general, it is believed that usually there are a number of different factors at work. These include exposure to harmful, *carcinogenic* substances (such as tobacco smoke), radiation, ultraviolet light, certain viruses and possibly, the presence in the body of *cancer genes* (oncogenes). The smoking of cigarettes is known to be a major cause of several types of malignant cancer.

capacitor a device used to store electric charge.

capillary (*plural* **capillaries**) in biology, any narrow or hairlike tube with a thin wall usually only one cell thick, e.g. blood capillary.

capillary action an activity related to SURFACE TENSION which results in liquid rising or falling in a narrow tube.

carbides chemical compounds of metals with carbon which usually produce very tough, hard substances, e.g. tungsten carbide. These substances are ideal for the manufacture of cutting tools which can be used on hard materials at high temperatures.

carbohydrates a large group of chemical compounds containing carbon, hydrogen and oxygen with the general formula $C \times (H_2O)$. The group includes the sugars, STARCH and CELLULOSE and carbohydrates play a vital role in the METABOLISM of all living ORGANISMS.

carbon a non-metallic element which occurs as diamond, graphite and carbon black. In industry it is used extensively as motor brushes, in steelmaking and in the manufacture of cathode-ray tubes (for televisions and computers). Carbon is unique in the enormous number of compounds it can form. It is present in all organic compounds and so in every living organism. It is made available to living things through atmospheric CARBON DIOXIDE which is taken up by plants during PHOTOSYNTHESIS (see CARBON CYCLE).

carbonaceous rocks SEDIMENTARY ROCKS that contain carbon derived from plant material. Examples are lignite, brown coal and true coal.

carbonates compounds containing carbon and oxygen as CO_3 e.g. CALCIUM CARBONATE.

carbon cycle the circulation of CARBON compounds in nature by the various life (METABOLIC) processes of many organisms. The main stages in the carbon cycle are:
1) Carbon dioxide present in air and water is taken up by green plants and some BACTERIA during PHOTOSYNTHESIS.

2) The carbon accumulated in plants is later released during the rotting and decay (decomposition) of the dead plant. It is also released by the decomposition of animals and bacteria that have eaten plants or other organisms. Through FOOD CHAINS, all organisms contain carbon which was originally taken up by green plants during photosynthesis.

3) Carbon, in the form of carbon dioxide, is released back into the environment as a waste product of the RESPIRATION of living organisms. It is also released by the burning of fossil fuels—coal, oil, gas and peat—which contain the remains of plants, and wood. The concentration of carbon dioxide in the atmosphere is increasing as huge areas of tropical forests are being destroyed, while the burning of fossil fuels is rising, hence there is less PHOTOSYNTHESIS to absorb the extra CO_2. This may be a factor in the small temperature rises throughout the world, known as the GREENHOUSE EFFECT. When there are high levels of carbon dioxide in the atmosphere, heat radiation from the sun tends to be reflected back to Earth rather than lost to space, leading to warming.

carbon dioxide (CO_2) a colourless GAS occurring in the atmosphere which solidifies at -78.5°C and is much used as a refrigerant. It is also used in carbonated drinks and, since fire does not burn in its presence, in fire extinguishers.

carbon monoxide a colourless gas formed during the incomplete burning of coke and similar fuels. It also occurs in the exhaust fumes of motor engines and is highly poisonous if breathed in. It combines with HAEMOGLOBIN in the BLOOD, and reduces the ability of blood to carry the oxygen necessary to support life processes.

carboxylic acids organic acids found in nature containing one or more carboxyl groups (acids) which have the formula -COOH.

carcinogen any substance that may predispose to CANCER.

cardiovascular system the organization of the heart, arteries and veins of the human body which form an almost closed system. With one or two exceptions, every part of the body needs a constant supply of blood in order to receive essential nutrients (foods) such as AMINO ACIDS and GLUCOSE, and oxygen. The exceptions are the dead structures such as the hair and nails (although the nail bed and hair roots need blood) and the cornea (the clear window of the eye) which is supplied by means of the tears.

carotenoids orange, red and yellow pigments found in some plants such as carrots and ripe tomatoes.

carnivore any animal that eats the flesh of other animals. The term can also refer to the group of mammals called *Carnivora* which includes bears, cats and dogs. Carnivorous plants are ones that trap and digest insects to obtain food.

carpel the female reproductive organ of a flower which consists of several structures called the stigma, style and ovary.

Cartesian co-ordinates a method of representing the position of a point in space, invented by the French mathematician and philosopher, René Descartes (1596–1650). For example, the point with Cartesian co-ordinates (5, 2) will be found by moving 5 units along the horizontal x-axis and 2 units up the vertical y-axis.

cartography the gathering of information, design and drawing of a new or revised map and also, anything connected with the study, presentation and use of maps.

catabolism any breakdown process of metabolism in which more complex substances of molecules are broken down into simpler ones.

catalyst a substance that increases the rate of a chemical reaction but can be recovered unchanged at the end of the reaction. Metal catalysts such as iron and platinum are widely used in industrial processes. Living organisms contain natural catalysts called ENZYMES which are essential in metabolic (life) processes.

cathode the negatively charge electrode of an electrochemical cell to which CATIONS travel and gain ELECTRONS (i.e. where REDUCTION occurs).

cathode rays a stream of ELECTRONS given off from the negative electrode (CATHODE) when electricity is passed through a VACUUM TUBE.

cation a positively charged ION formed by an ATOM or group of atoms that has lost one or more ELECTRONS.

celestial body any of the stars or planets that can be studied by the methods of ASTRONOMY.

celestial equator the circle created by the meeting of a projection into space of the plane of the Earth's equator, with the CELESTIAL SPHERE. It marks the boundary between the southern and northern hemispheres.

celestial poles the points at which imaginary northerly and southerly extensions of the Earth's axis (through its POLES) meet the CELESTIAL SPHERE.

celestial sphere an imaginary sphere which places an observer on Earth at its centre and all the stars, whatever their true distance, on the inner surface of the sphere.

cell the basic unit or building block of all living organisms. Usually cells are only visible with the aid of a microscope and some organisms consists of only one cell and are called UNICELLULAR (see BACTERIUM). Cells also merge to form tissues or colonies. There are two types of cell, PROCARYOTE and EUCARYOTE, the most primitive ones being procaryotes.

cellulose a naturally-occurring CARBOHYDRATE substance which occurs widely as cell walls in plants. It is a polysaccharide (see SACCHARIDE) consisting of chains of glucose (sugar) units, and gives strength to the structure of a plant. It occurs in wood, cotton grass and other natural plant fibres and is used in the manufacture of paper, plastics and explosives.

Celsius scale or **Centigrade scale (C)** a temperature scale with a freezing point of 0° and boiling point of 100°, devised by the Swedish astronomer Anders Celsius (1701–1744).

central nervous system the part of the nervous system that receives, decodes and sends out all nerve signals. In VERTEBRATE animals, it consists of the BRAIN and SPINAL CORD.

centre of gravity gravity acts upon all parts of a body and the sum effect of these forces acts through a single point, called the centre of gravity.

centrifugal and centripetal forces are those which are set up when a body is spinning fast around a central point, and cannot move out or away from its position. The centripetal force is tending to pull the object or body in towards the centre while the centrifugal force is acting in direct and equal opposition. The centrifugal force is a reaction to the centripetal force and the result is that the object stays at the outer limit of the circle of rotation. These forces are often used to create exciting fairground rides.

CFC an abbreviation for chlorofluorocarbon, a chemical widely used in manufacturing processes, especially refrigeration, which reacts with and destroys ozone, causing a thinning of the OZONE LAYER.

chain reaction an overall chemical reaction which takes place in many or several stages. The compounds, products or molecules formed at each stage are the trigger for the next step in the reaction. The reaction slows down and stops when all the products in the final stage have been used up.

chalk a white, soft, fine-grained limestone which is porous (allowing air and water to penetrate easily) and is made from CALCIUM CARBONATE and the skeletons of fossil MICRO-ORGANISMS. Chalk deposits are widespread in much of north-west Europe.

Charles' law a law that states that at constant pressure, the volume of a gas varies directly with the temperature. Hence if the temperature is doubled then the volume will double also. (See also GAS LAWS).

chelation a reaction between a metal ion and an organic molecule which ties up

the metal into a closed stable ring structure. It is a process which occurs naturally in the soil, removing metals which might be harmful to plants. The principle
is applied to domestic products such as shampoos and detergents – chelating
agents are added to soften water by locking up calcium, iron and magnesium
ions.

chemical equation the writing of a chemical reaction using symbols to represent the ATOMS and MOLECULES.

chemistry the study of the composition of substances, their effect upon one
another and the changes which they undergo. The three main branches of
chemistry are ORGANIC, INORGANIC and PHYSICAL chemistry.

chemotherapy the use of toxic (poisonous) chemical substances in the form of
drugs to treat diseases. The chemicals are targeted against abnormal, cancerous
tissue or invading, disease-causing MICRO-ORGANISMS in conditions where a person's life is at risk.

china clay a clay, mainly composed of kaolinite, which is formed from the alteration of granite rocks due to weathering. It is extracted using high pressure water jets and is used in many industries including ceramics, paper and pharmaceuticals.

chi-squared test a test in STATISTICS used to determine how well data (results)
obtained from an experiment (the observed data) fits in with the data expected
to occur by chance. The chi-squared test is a simple method of checking that
the experimental results are significant and have not just arisen from chance
events.

chitin a naturally-occurring substance which is a HYDROCARBON and similar to CEL-
LULOSE, but contains NITROGEN. It forms the skeleton of many INVERTEBRATE animals, e.g. insects.

chlorine a chemical substance that exists as a yellow/green gas which has a
harmful effect if breathed in, being an irritant and causing choking. It occurs
widely in the form of chloride, the commonest being salt (NaCl), or sodium
chloride. Chlorine is used in the production of bleaches, disinfectant and hydrochloric acid. Also, it is used in the production of organic chemicals, e.g. carbon tetrachloride, PVC, plastics and solvents.

chlorofluorocarbons see **cfc, ozone layer**.

chlorophyll the green pigment of plant cells, contained in ORGANELLES called
CHLOROPLASTS which are found in some plant cells, especially the leaves of green
plants. Chlorophyll is essential for PHOTOSYNTHESIS as it traps energy from sunlight and uses it to split water MOLECULES into HYDROGEN and OXYGEN.

chloroplast an ORGANELLE found within the cells of green plants and ALGAE, where PHOTOSYNTHESIS takes place. The chloroplasts contain the enzymes necessary for the CALVIN CYCLE and CHLOROPHYLL, the essential green pigment which traps light energy from the sun.

cholesterol a naturally-occurring FAT that is abundant in animal cells, occurring as insoluble (i.e. does not dissolve in water) molecules of saturated fat. In mammals, cholesterol is made from saturated fatty acids in the liver and transported in the blood by carrier molecules. Special receptor sites on the surface of cells recognize and mop up the carrier molecules and the cholesterol is taken up in this way. This mechanism of uptake helps regulate the levels of cholesterol in the blood. If a person's diet is high in cholesterol, the number of receptors on the cells decreases and there is less uptake of cholesterol by cells. This leads to an increase in the levels in the blood, which can be harmful as the excess is deposited in the walls of arteries. Eventually, this leads to ARTERIOSCLEROSIS, a narrowing and hardening of the arteries. Although cholesterol can cause health problems, and is a subject of concern for people in western countries, it should be remembered that it is an essential substance in the body. It is a major constituent of the surrounding layer (MEMBRANE) of cells, STEROID HORMONES and BILE salts.

chromatid one of a pair of side-by-side replica CHROMOSOMES, joined together at a point called the CENTROMERE, which are produced during the replicating or copying of DNA during cell division.

chromatography a method of analysis used to isolate the different parts or constituents contained in a solution or gas. It works by exploiting the fact that the various molecules are held together by bonds of different strengths. Due to this the constituents separate out (and are collected on absorbent material) at different rates and the amounts and concentrations can then be analysed.

chromosome a structure found within the NUCLEUS of eucaryotic (see EUCARYOTE) cells, consisting of DNA and PROTEINS. The chromosome takes part in three processes connected with cell division.

Replication—this ensures the correct copying or duplication of the DNA.

Segregation—this ensures that the newly-copied or replicated chromosomes will separate and that each will become part of a daughter call. This is a very important function, largely dependent upon the CENTROMERE.

Expression—this ensures that the GENES (DNA) present on the chromosome are correctly copied to preserve the particular information which they code for.

Procaryotic cells (see PROCARYOTE), have a single circular chromosome which is not contained within a nucleus. A species has a characteristic number and pattern of genes and chromosomes. For example, yeast cells contain 16 pairs of chromosomes and human beings have 46 pairs.

Chromosomes can only be seen under the microscope during MITOSIS or MEIOSIS when they become shorter, fatter and thicker before separating.

cilia (*singular*, **cilium**) a fine, thread-like hair projecting from the surface of a cell. Cilia beat to create currents of liquid over the cell surface or to cause the cell to move.

circuit a pathway that, when complete, allows electric charge to flow through it. There are two types of circuit, PARALLEL and SERIES. In series circuits the same amount of current flows through all the component parts, but there is a varying drop in POTENTIAL DIFFERENCE through each component. In parallel circuits, a different amount of current flows through each component part, but the same drop in potential difference occurs across each component of the circuit.

cirrocumulus a type of high level cloud formed in stable air as sheets or layers with ripples and waves.

cirrostratus a type of cloud of white or near transparent sheets which may cover the sky, producing a rainbow or white ring around the SUN or MOON due to refraction by ice crystals.

cirrus a high level cloud forming narrow bands or streaks.

citric acid a naturally occurring chemical substance which is a tricarboxylic acid, found in many fruits, especially citrus fruits such as lemons. It has the formula $C_3H_5O(COOH)_3$ and is used commercially as a food flavouring and in the production of fizzy drinks and salts for indigestion.

citric acid cycle a complicated series of biochemical reactions controlled by ENZYMES. The reactions take place within living cells, producing energy, and is essential in the final stages of the OXIDATION of CARBOHYDRATES and fats. It is also involved in the production of some AMINO ACIDS.

clathrate a chemical compound in which MOLECULES of one type are enclosed in the structure of another different type of molecule.

clay a type of mud or sediment that is very fine-grained and is made up of clay minerals which chemically are hydrous (water-containing) ALUMINIUM SILICATES. Clay has numerous uses in industry including the making of ceramics and bricks, rubber, plastics, paints, as fillers in paper manufacture and in drilling muds.

cleavage the tendency of minerals and rocks to shear and split along particular planes, depending upon their internal structure.

climatic zones regions of the Earth in which there is a distinct pattern of climate, each region approximately the same as belts of LATITUDE. There are eight climatic zones in each hemisphere of the Earth. Beginning at the polar ice cap in the northern hemisphere, there is a *boreal* (northern) zone with a range of temperatures, two belts of humid, temperate conditions, two sub-tropical, arid or semi-arid zones and, finally, the humid tropical belt near the Equator.

clone an organism, micro-organism or cell derived from one individual by asexual reproduction. All the organisms or clones have the same genetic make-up (GENOTYPE). Cuttings taken from plants and grown into new individuals are clones which are identical to the parent plant.

cloning is an artificial process engineered by man and used in the breeding of plants (e.g. cuttings). in February 1997, scientists in Scotland claimed they had successfully produced a ewe, which they called Dolly, that had been cloned from another adult ewe. However, it was later conceded that there was a very remote possibility that Dolly might have been cloned from foetal cells cultivating in the ewe's bloodstream.

closed-chain compounds in ORGANIC CHEMISTRY in which carbon atoms are bonded together to form a ring, e.g. BENZENE. Therefore, closed-chain compounds are called ring or cyclic compounds. (*See also* HETEROCYCLIC COMPOUNDS).

cloud water droplets or ice formed by condensation of moisture in rising warm air. Rising on convection currents, the warm moist air reaches the higher, cooler conditions and condenses. Clouds are classified on their form and shape into three major groups—cumulus (heap), stratus (sheet) and cirrus (fibrous). These are further sub-divided according to such features as transparency, type of growth and arrangement.

cloud chamber or Wilson cloud chamber an instrument that makes visible a stream of charged particles moving through a gas.

coal see fossil fuels

coal tar one by-product from the breakdown of coal by heat in the absence of air (i.e. carbonization, a process for the production of coke, town gas and smokeless fuel). Depending upon the temperature of carbonization, coal tar contains a cocktail of chemicals including BENZENE, toluene, PHENOLS and pyridine.

coefficient the constant numerical (number) part of a term in an equation in algebra. For example, in the equation $5x - 6xy + y = 0$, the coefficient of x is 5, that of xy is -6, and that of y is 1. In some mathematical expressions in which the coefficients are unknown, these can be represented by letters and their values calculated using a particular formula.

cohesion the attraction between molecules in a liquid that allows thin films and drops to form.

cold front in weather forecasting, the line defining the front of a mass of cold air that will cut under warm air lying ahead of it. The weather changes that may occur include a fall in the temperature and a squally wind which may change its direction.

collagen an important fibrous PROTEIN that forms almost one third of the total body protein in mammals, and is found in tendons, bone, skin and cartilage.

colligative property a property of a solution that depends entirely upon the concentration of the dissolved particles rather than their nature. Colligative properties are important in determining or forming osmotic pressure (see OSMOSIS), vapour pressure or the freezing point of a solution.

collision in physics, an interaction or meeting between particles in which MO-MENTUM (movement) is maintained. The collision is called ELASTIC if the KINETIC ENERGY of the particles is kept and none is lost. However, it is called INELASTIC if the kinetic energy is not maintained. In nuclear physics, collision usually means a closeness of particles that enables interactions between the forces that surround them to take place. An actual collision would result in capture of a particle.

colloid a substance forming particles in a solution varying in size from fine ones in a true solution to coarser ones in a SUSPENSION. The particles are charged and can be subjected to ELECTROPHORESIS, and measure 10^{-4} to 10^{-6} mm. Many substances e.g. vegetable fibres, rubber and proteins occur naturally or are at their most stable in the colloidal state. Different types of colloids are recognized—gels, emulsions, sols, aerosols and foams.

colony a group of individuals of the same species living together and depending upon one another to a greater or lesser extent. Sometimes the individual organisms are actually joined, as in corals, and live as one larger unit. In other species, e.g. some insects, the individuals remain separate but exist in a highly organized system and may carry out different tasks. Also, a colony may be a BACTERIUM or yeast growing from a parent cell on a supply of food.

colorimeter an apparatus used to measure the hue, brightness and purity of a colour. In colorimetric analysis, the colour of a solution is analysed in comparison to a standard solution in which the type and amounts of the particles present are known.

comet a small body moving in an irregular orbit around the Sun and composed of rock, dust, gas and frozen ices (of carbon dioxide, carbon monoxide, water and formaldehyde). A comet is formed of a head, consisting of an irregularly shaped nucleus a few kilometres across and a coma, which is a layer of gas. The other part of a comet is the tail, consisting of gas and dust which trails out from the head and is formed when the orbit takes the comet close to the Sun. When the orbit takes the comet a long distance from the Sun, it cannot be seen from Earth. When it approaches close to the Sun it appears to flare up and look bright and becomes visible from Earth. One complete orbit of the Sun is called the period of the comet. The shortest period belongs to Ericke's comet and is only $3^1/_2$ years, but others have much longer periods. The most familiar comet, which has excited interest for a long time, is Halley's comet which appears every 76 years and is next due in 2062.

commutator the part in a motor or ARMATURE that makes contact with the carbon brushes to carry current.

common logarithm see **logarithm**.

complementary angles two angles totalling 90°. With such angles one is the complement of the other.

component a technique used in physics in which a single force can be split into component forces. When acting together, the component forces produce an effect which is the same as the resultant force acting alone. Used in reverse, it allows the size and direction of a force to be determined from two components.

compound in chemistry, when two or more elements are combined in a substance in definite proportions, in the form of MOLECULES held together by chemical bonds.

concentration the quantity of a substance (called the SOLUTE) dissolved in a fixed amount of liquid (the SOLVENT) to form a solution. Concentration is measured in moles per litre (mol 1-1).

concentric a term used in mathematics to describe geometrical figures that share a common centre.

condensation the process by which a substance changes from being in the form of a gas to the form of liquid. KINETIC ENERGY is lost during the process. In meteorology, condensation occurs when warm, moist air is cooled below its

DEWPOINT and particles (such as dust or ions) provide surfaces on which water droplets are formed.

In chemistry, condensation is the reaction of one molecule with another, and the elimination of a simple molecule such as water or an alcohol.

condenser in chemistry, a piece of apparatus used to condense vapours which is usually in the form of a tube or tubes, surrounded and cooled by a water-filled jacket.

In physics, a condenser is a large lens or mirror that collects and directs light in projection equipment or apparatus. The light is directed onto the transparency or other object needing to be focused by the projection lens.

conductance the reciprocal or opposite of RESISTANCE in an electrical circuit. This used to be measured in units called reciprocal Ohms (mho) but in the SI system, the unit used is the siemen (s).

conduction conduction of heat through a solid, or the flow of electrical charge through a substance. Heat moves through a material due to the vibration of the molecules of which it is composed. When the material is heated, the molecules vibrate rapidly and knock into neighbouring molecules which transfers the heat (or thermal energy) along the material. Transfer of heat energy by conduction is always from a high temperature to a region of lower temperature. Metals, such as copper and aluminium, are the best conductors of heat. Also, all metals and carbon (in the form of GRAPHITE) are good conductors of electric charge as they contain electrons which are free to move about or flow, and thus carry and transfer energy as a current. The electrons around the outside of the atoms are only loosely held and can move easily. A poor conductor of heat and electricity is called an INSULATOR, e.g. plastic, cork, glass and air and most non-metals. They have electrons which are not free to move and are not able to conduct electric charge.

conductivity the reciprocal or opposite of RESISTANCE in an electric circuit. The units of measurement are siemens per metre.

connective tissue a common tissue found in animals that is further divided into various kinds, depending upon the amounts of materials it contains and the function it performs. It is usually composed of a non-living core or matrix containing various fibres in which are spread a number of different types of CELLS. Examples of connective tissue are blood, lymph, bone, cartilage and adipose (fatty) tissue.

conservation of energy, law of thermodynamics see **energy**.

constant in mathematical equations or algebraic expressions, any quantity that remains the same.

constellation a group of stars which are placed together and often represented as an outline or picture. They are given a name but there is no scientific basis for the grouping. Many were named by ancient astronomers and examples include the Great Bear (Ursa Major), the Bull (Taurus), the Hunter (Orion) and the Pole or North Star.

continental crust the Earth's crust that lies beneath the continents and the continental shelves. It is usually 30–40 km or 19–25 miles thick but this increases to about 70 km or 44 miles beneath areas of mountain building. The crust consists of two layers made up of different materials and which vary slightly in density. The base is marked by a level where composition and density change greatly, known as the Mohorovicic discontinuity.

continental drift an idea or concept in geology, devised by a Germany geophysicist Alfred Wegener (1880–1930). He suggested that 200 million years ago the Earth's surface had one large single continent called Pangaea, which broke apart to form the present continents. A way of explaining how such huge land masses move, is provided by the modern area of scientific geological study called plate tectonics. The Earth's crust is formed of a number of sections or plates. These are believed to float on a partially molten layer, called the lower mantle, which lies between them and the Earth's core.

continental shelf the area of the Earth's surface lying between the shoreline and the top of the continental slope, which is covered by shallower seas and supports an abundance of living organisms. At the top of the continental slope, the gradient of the surface becomes much steeper, at a depth of approximately 150 metres (490 feet), leading down to the largely unknown regions of the ocean floor.

contour a line connecting points on a surface that are at the same height above the reference surface (the *datum*). On a map, a contour line links points at the same height above sea level.

convection in a liquid or gas that is free to move, convection is the process that transfers heat from one part to another. It only occurs if the cooler area is above the hotter part of the liquid or gas. As water in a tank is heated, the hot water rises and the cooler water sinks, establishing a convection current. This system is used in the hot water heating systems in houses. A similar effect occurs with gases and when air is warmed, setting up convection currents which rise up from a hot land surface. On a larger scale, onshore and offshore winds are created in coastal regions. An everyday example that can be easily felt with the hand is the hot air that rises from a heated radiator.

copper (Cu) a highly malleable (i.e. easily worked), red-brown metal which is ductile (see DUCTILITY) and has many uses. It occurs as native copper (as the metal itself, often with silver, lead and other metals) and in a variety of mineral forms including malachite, bornite and chalcopyrite. The ores are concentrated and copper is extracted by smelting and refining by ELECTROLYSIS. Copper has been an important metal for thousands of years in its alloys, brass and bronze, and it is now used in coins as an alloy (with nickel).

Since copper is an excellent conductor of electricity, it is widely used in the electrical industry. It is also used as pipes for plumbing and in paints, pigments, printing and fungicides (chemical solutions which destroy fungal growths).

copolymer a POLYMER compound, formed by the polymerization of two or more MONOMERS, a number of which are important in industry.

coral reef a hard bank built up from the carbonate skeletons of corals that live in COLONIES (and algae). There are several forms of reef, including BARRIER REEFS, fringing reefs, which are attached to the coast, and atolls, where a reef encloses a lagoon. In order for a reef to grow, certain conditions must be present. The maximum depth of water must not be greater than 10 metres or 33 feet, it must be clear and not contain mud washed from the land and it must be within a certain temperature range. Also, the water must be of normal salinity (saltiness).

Coriolis force a theoretical force used when calculating the movement of particles in relation to a spinning or rotating body, such as air passing over the Earth's surface.

corona in METEOROLOGY, coloured rings, seen through thin cloud caused by water droplets diffracting light rays. Also, a term in physics involving air on the surface of a conductor.

coronary artery one of a pair of large blood vessels which branch from the AORTA, the major artery which carries blood away from the heart. These carry blood containing oxygen to the heart muscle.

corrosion the gradual wearing away of solids, especially metals and alloys by chemical attack, due to acids or alkalis carried in air or water.

corundum a hard mineral oxide of aluminium, with the chemical formula Al_2O_3, which occurs as gemstones, i.e. sapphires (blue) and rubies (red). As it is very hard it is used in industry in grinding and abrasive processes.

cosmic rays rays of radiation coming from various sources in space such as the Sun and solar wind.

cosmology the scientific study of the nature of the universe including its beginnings (origin), structure and development (evolution).

coulomb a unit of electric charge (C) which is the amount of electricity carried through a particular distance in one second by a current of one AMPERE (A). Hence: charge (C) = current (A) x Time (s).

covalent bond the joining of two atoms due to the equal sharing of their electrons. If only one pair of electrons are shared, this is known as a single covalent bond, if two, a double and if three, a triple covalent bond. The bonds are usually very strong and stable.

CPU (Central Processing Unit) the part of a computer that accepts and processes information.

cracking an industrial process in which large complicated molecules are broken down into smaller ones. It usually involves the use of heat but pressure and CATALYSTS are also used. Cracking is most commonly used in the petrochemical industry.

crucible a heat-resistant container used in scientific and industrial laboratories, for carrying out experiments often involving the use of heat.

crude oil petroleum in its natural, unrefined state.

cryogenics the scientific study of the behaviour of materials and substances at very low temperatures. Usually liquefied gases (cryogens) are used.

crystal a solid material with a regular ordered structure, having faces, which are usually flat, on several sides.

cube see polyhedron.

cumulonimbus large, bulging clouds that reach great heights, the upper parts forming anvil shapes or plumes. The base is dark and usually produces rain.

cumulus well-defined clouds as separate, bulging masses often with a flat base which is dark.

current the flow of charge in an electrical circuit. Current (I) is measured in AMPERES (A).

cybernetics a science developed by an American mathematician, Norbert Wiener (1894–1964) which is the study of organization, regulation and communication within control systems.

cyclone see depression.

cyclothem a feature of some SEDIMENTARY ROCKS in which a set of deposits is repeated in a cycle or sequence. It is a typical feature of coal-bearing rocks.

cyclotron a machine that produces high-speed charged particles for use in nuclear experiments.

cytogenesis the formation and development of the CELLS of living organisms.

cytogenetics the scientific study of the structure and behaviour of chromosomes and the way in which features of parent organisms are passed on to the offspring.

cytology the study of all aspects of the CELLS of living organisms including their structure and the activities which they carry out.

D

Dalton's atomic theory a milestone in the development of chemistry devised by the English chemist and physicist John Dalton (1766–1844). He put forward the theory that all matter is made up of particles (atoms), which are identical throughout one element or substance, and that chemical reaction occurs through the attraction between atoms.

Dalton's law of partial pressures in a mixture of gases, the total pressure is the sum of the partial pressures of all the gases present. The partial pressure is the pressure a gas in a mixture would exert if it alone occupied the space of volume.

dark nebulae clouds of thick gas and dust common throughout the MILKY WAY and other GALAXIES.

dark reaction a stage in PHOTOSYNTHESIS which is not directly dependent on light. It takes place within the CHLOROPLASTS of the cells of green plants.

Darwinism the theory of evolution devised by the British naturalist, Charles Robert Darwin (1809 –1882), in order to explain the great variety of plants and animals which exist on Earth. He arrived at his theories as a result of a five year voyage around the world (the voyage of the *Beagle*) and when he returned to England in 1859, he published a scientific paper with the title *Origin of Species*. Darwin proposed that some individuals in a species are more successful than others and hence better adapted to their environment. They are more likely to reproduce successfully and some of the characteristics which enable them to do so are inherited by their offspring. Hence these characteristics eventually become more widespread within a species. Gradually, new species can emerge that will adapt to new environments. Old species, which are no longer suited to the surrounding environment, will eventually die out.

decibel a unit for measuring the level of sound which has the symbol dB.

decimal numbers a structured system of numbers based on ten and the most commonly used number system. The decimal point is the dot that divides the number's whole part from the fractional part (i.e. that which is less than one). However, numbers need not contain a decimal point; 789 is also a decimal number. A decimal number itself is less than one, e.g., 0.789 and 0.00987. The value given to a digit within a number depends upon its position in the number. With the decimal system, each column has ten times the value of the column on its immediate right. Therefore the number 7891 is:

seven 1000s (10^3), eight 100s (10^2) nine 10s (10^1) and one 1 (10^0)

    7   8   9   1

A common fraction such as $1/4$ can be changed into decimal form by dividing the 1 by 4 to give 0.25.

decomposer any organism that breaks down dead organic material (i.e., the material derived from living things), which may be remains of plants or animals or waste (such as animal dung). Decomposers get the energy they need from this process and change the material into a simpler form. Earthworms are a good example of an animal decomposer but most are bacteria and fungi.

decomposition the breakdown of a substance from a more complicated form into a simpler one, which occurs in both chemical and biological processes.

deformation processes that rocks may undergo over a long period of time during which they are changed from their original state. Evidence for these processes is obtained from the geological study of rocks. Over the millions of years of the Earth's history, there have been periods of mountain building and pushing up of very hot, molten material from beneath the crust. This has resulted in the cracking (faults), folding (folds) and disturbance by igneous intrusion, of the overlying rocks.

degredation the breakdown of complicated molecules into more simple ones. In physics, it refers to a loss of KINETIC ENERGY due to collision of particles.

dehydration in chemistry, the removal of a water molecule from a compound or more complex molecule by the action of heat. A CATALYST or chemical agent often acts as a dehydrating agent, e.g. SULPHURIC ACID. In medicine, dehydration is the loss of a large amount of water from the cells and tissues of the body which is often dangerous.

deliquescence the situation in which a substance picks up water from the air and may eventually turn to liquid. It happens because the substance has a very low water VAPOUR PRESSURE, lower than that of the surrounding air, so water is absorbed.

delta a roughly triangular area of mud, clay, sand or silt (sediment) formed at the mouth of a river. It is caused by a current as in a river, which is laden with sediment meeting and entering a large body of water such as the sea or a lake. There is a slowing of the speed of the current and so its ability to carry sediment is reduced. Therefore much of the material is deposited on entering the lake or sea. The shape of the delta depends upon various factors including the volume of water, speed of current, sediment load, tides and climate. Well-known examples are the deltas of the Nile and Mississippi.

denaturation the breaking of the weak bonds that hold protein molecules together, usually caused by extreme heat or the action of acids or alkalis.

dendrochronology the science of determining age through studying the annual tree rings which are laid down every year during the life and growth of a tree.

denominator the number below the line in a VULGAR FRACTION, e.g. 3 in $^2/_3$.

density is the mass over volume of a substance or gas and is measured in units of kilograms per cubic metre ($kg\ m^{-3}$). It can be calculated using the equation:

$$density\ (d) = mass\ (m)/volume\ (v)$$

Therefore a certain volume of gas will have a greater density if it is put into a small container than if it is placed in a large one.

denudation the "stripping bare" of a land surface by the forces of climate and weather, known as WEATHERING and EROSION.

deoxyribonucleic acid see **DNA**.

dependent variable in a mathematical expression, the quantity with a value which depends upon the other INDEPENDENT VARIABLES. For example, in the equation $y = 6x + 3$, y is the dependent variable as the value of y depends upon that inserted for x which is the independent variable.

deposition the laying down of sediments (clay, mud, silt, sand, etc. and also mineral veins in rocks).

depression in a weather system, an area of low pressure, also called a cyclone, which has particular circular patterns of wind flow. These winds flow anticlockwise in the northern hemisphere and clockwise in the southern hemisphere. The weather associated with a depression is usually unsettled and stormy, with wind and precipitation (rain, hail, sleet or snow).

desiccant a substance that absorbs moisture and is used to dry out materials. Examples include calcium chloride and silica gel.

detergent a soluble substance that acts as a cleansing agent and is especially good at the removal of grease and oils which are both HYDROCARBONS. The detergent molecules have a dual action in that one part of the molecules is water-hating (hydrophobic), while the other is water-loving (hydrophilic), having an affinity for water. The hydrophobic part interacts with the hydrocarbons (making up the grease or oil), and the hydrophilic part reacts with water. This dual action brings the water and grease together so that the mixture can be rinsed away with clean water.

dew condensation of water vapour in the air producing water droplets or moisture. Dew is formed when the temperature falls below the DEWPOINT and the air is saturated and can no longer hold all the water vapour. (Warm air can hold more water vapour than cold air). The water condenses out of the air as droplets on cool surfaces.

dewpoint the temperature at which air becomes saturated with water vapour and deposits DEW.

diagenesis the changes that occur in a sediment (e.g. sand, silt, clay or mud) after deposition. These occur at ordinary pressures and temperatures and are generally non-dramatic but are responsible, over a long period of geological time, for transforming the sediment into rock.

dialysis a method for separating small molecules from larger ones in a solution. Dialysis occurs in the kidneys of all vertebrate animals and is the process which cleans the blood of the waste products of METABOLISM.

dichotomy dividing into two equal parts, e.g., in botany, where the growing point of a plant divides to produce two further, equal growing points that also divide after more growth. Also, in astronomy, when a planet or moon is seen half lit-up.

dielectric a substance, gas or liquid that does not conduct an electric current and so is an INSULATOR. Dielectrics can be used for capacitors (see CAPACITANCE), cables and terminals.

differentiation a procedure used in CALCULUS for finding the derivative of a function. In geology, the process by which several types of (IGNEOUS) rock are produced from one 'parent' MAGMA (hot, liquid rock). As particular minerals crystallize out from the magma as it cools, the composition of the remainder is altered. Hence a different group of minerals crystallizes out from the magma which is left, forming another kind of rock.

diffraction the bending of waves, whether water, light, sound or electromagnetic, around an obstacle such as the straight edge of a barrier. The diffraction of all waves around such an object can be detected by a change in the shape of the wave front. There is no change in the wave frequency or wavelength, and so the speed of the wave stays the same provided that the surrounding properties remain constant.

diffusion the natural process by which molecules will disperse evenly throughout a particular substance. The molecules always travel down their concentration gradients, that is they move from a region where they are highly concentrated to one where they have a lower concentration within that substance. Diffusion occurs in gases and liquids and, depending upon the size of the molecules, across the membranes of CELLS and ORGANELLES inside cells.

diffusivity the rate at which heat diffuses through a material measured in square metres per second.

diode a device that will allow current to flow in only one direction. A diode valve has a solid surface that, when heated, gives off electrons and hence becomes the negatively charged plate (CATHODE). The current travels towards a positively charged plate, the anode, which is connected to the positive terminal of a power source such as a battery.

diploid a term used to describe a cell having two of each CHROMOSOME in its nucleus. These are called HOMOLOGOUS pairs of chromosomes with each one having a similar distinctive shape. Humans are diploid, with each cell of the body, except GAMETES (or sex cells) having 46 chromosomes, that is 22 pairs of homologous chromosomes (known as autosomes) and one pair of sex chromosomes.

dipole one of two equal and opposite charges (electric dipole) or magnetic poles (magnetic dipole) separated by a short distance.

direct current (DC) an electric current flowing in one direction only.

disintegration the process by which one or more particles is given off from the nucleus of an atom, especially in the decay of radioactive materials.

displacement the calculation to determine the distance of a change in position of an object by relating the final position to the position it had at first. Displacement is measured in metres and is not related to the path the object travelled to reach its final position.

dissociation the process by which a compound breaks up into smaller MOLECULES, or IONS and ATOMS.

dissolution the dissolving of a substance in a liquid to form a solution in which all the material is evenly distributed (a homogenous solution).

distance in physics, the measurement of how far an object has travelled along a particular path. In mathematics, distance is the length of a line needed to join particular points and is similar to displacement. The units used for measurement are metres.

distillation the separation of a liquid solution into its various parts (or components), by heating it until it becomes a vapour and then cooling the vapour so that it condenses and can be collected. The member parts of a solution can be collected at different times because each has its own BOILING point at which it turns into a vapour.

diurnal daily, i.e., happening every 24 hours. Many animals have a diurnal rhythm of behaviour involving sleeping, feeding and other activities.

divergent junction or constructive boundary in geology, a boundary between two of the Earth's crustal plates that are moving apart because new material is being made. Major activities are associated with these boundaries, the most spectacular of which are earthquakes and volcanoes.

DNA (*abbreviation for* **deoxyribonucleic acid**) strands of material present in the nucleus of the cells of all living organisms. It occurs as two long strands twisted round one another to form a structure called a double helix. The two strands are linked by bridges of material called nucleotides. The DNA strands carry the GENES which are responsible for all the characteristics of the organism and are passed from parents to offspring.

doldrums fairly calm areas of ocean around the Equator which experience low pressure and light winds.

dominance in genetics, where one particular ALLELE or GENE for a certain characteristic is dominant over another. A familiar example is eye colour in humans where brown is dominant over blue.

Doppler effect a change in the apparent frequency of a wave (such as light and sound) experienced by a person if the source of the wave is moving. The Doppler effect is responsible for an ambulance siren having a higher pitch as it approaches you and a lower pitch as it goes away into the distance.

double decomposition a chemical reaction between two substances that result in their breakdown and two new substances being formed from the parts, e.g.,

$$AB + XY \rightarrow AX + BY.$$

drowned valleys river valleys that have been flooded due to a rise in sea level, often following the melting of ice after the last Ice Age.

dry ice the solid form of carbon dioxide (CO_2) which is used in refrigeration.

dry valley a valley once carved out by a river but now dry.

ductility the property of metals or alloys which allows them to be drawn out into a wire and to keep their strength when their shape is changed.

dune a collection of sediment, usually sand, which also moves under the influence of the wind. Dunes in the Sahara reach a height of 300 metres or 1000 feet.

duodenum the first length of the small intestine or gut of vertebrate animals, which connects the stomach to the next part (the ileum).

dyke an INTRUSION of IGNEOUS rock forced up from the hot molten material beneath the Earth's crust and which has become solid. Dykes are usually vertical or nearly so and cut through the rocks into which they are forced.

dynamics a branch of physics dealing with the movement of objects and the forces acting upon them.

dynamo a machine that makes (generates) electric currents using a spinning or rotating coil as a conductor and powerful magnets to make a MAGNETIC FIELD.

E

Earth the Earth is one of a group of nine planets which orbit a star that we call the sun. It occurs in one small part of the galaxy (also called the milky way) which is referred to as the solar system. The Earth is the fifth largest of the planets and the third nearest to the Sun. Most of the Earth's surface is covered by the oceans—about two-thirds compared to one third which is land. A layer of air, the atmosphere, surrounds the Earth which is sub-divided into a number of different regions according to height above the surface. The Earth, as viewed from space, is like a blue ball or sphere which is flattened at both the poles. At the equator, the radius of the Earth is 6378 km or 3963 miles.

The Earth itself is composed of four main layers; an outer *crust* overlying a *mantle*, followed by an *outer core* and *inner core*. The rocky crust varies in thickness, being about 40 km thick under the continents and 8 km under the oceans. It 'floats' on the mantle which is made up of extremely hot rock and is about 2900 km thick. The core is so hot that it is part liquid, with a temperature in the region of 6500 km. The outer core is believed to be about 2250 km thick and is thought to be more liquid than the inner core which is solid and composed mainly of iron and nickel.

earthquake the often violent movement of the earth along a fault or fault zone due to tectonic upheaval caused by the sudden release of stress that has built up over some time. The focus is the point at which the earthquake originates within the earth and earthquakes are classified by depth of focus: shallow—less than 70 km; intermediate —70-300 km; deep—more than 300 km (see EPICENTRE and RICHTER SCALE). Attempts to predict earthquakes rely upon measurement of stress increases, but can so far be related only to an increasing risk of activity.

ECG (*abbreviation for* **electrocardiograph**) equipment used to record the current and voltage associated with contractions of the heart.

eclipse the total or partial disappearance of one astronomical body by passing into the shadow of another. A solar eclipse occurs when the new moon passes between the earth and the sun. A lunar eclipse occurs when the moon moves into the shadow of the earth, i.e., the earth is situated between the sun and the moon.

ecology the study of the relationship between plants and animals and their environment. Ecology is also known as bionomics and is concerned with, for example, predator-prey relationships, population dynamics and competition between species.

ecosystem an ecological community that includes all organisms which occur naturally within a specific area.

Einstein, Albert (1879-1955) a German-born American physicist and mathematician who formulated the RELATIVITY theories in 1905 and 1915 and carried out important investigations in THERMODYNAMICS and radiation physics. In 1921, he received the Nobel Prize in physics. He became a professor of mathematics at Princeton University, New Jersey, in 1933 after fleeing Nazi Germany.

elastic limit (see HOOKE'S LAW) the point beyond which a body e.g., a metal spring will deform when stress is applied rather than returning to its original state. In the case of a spring, the elastic limit is exceeded when it is pulled out of shape instead of springing back to its usual form.

elasticity a property of any material that will stretch when forces are applied to it and recover when the forces are relaxed. To stretch a spring or any other elastic material, equal but opposite forces (opposite in the sense of direction) must be applied to two areas of the material. All materials are elastic to some extent and will obey HOOKE'S LAW if the forces applied do not cause permanent deformation.

elastomer usually a synthetic material which has properties similar to natural rubber e.g., a capacity to return to its original state after being extended. In addition to rubber, elastomers include synthetic rubbers (such as polybutadienes) and other materials.

electric current a flow of electric charge through a conductor. The charge may be carried by means of electrons, ions or holes. Hole is the term for the absence of an electron in the electronic structure of a material. The movement of an electron in to the hole creates new holes and therefore "conduction by holes" (see also CURRENT).

electric field the invisible force that always surrounds any charged particle. When one charged particle is within the proximity of the electric field of another charged particle, each will feel and exert a force. If the charged particles are opposite (unlike), then they will attract since opposite charges attract. However, two particles with the same charge, i.e., both positive or both negative, will repel each other. An electric field is diagrammatically represented by lines along which a free positive charge would theoretically move, and so the arrows will always point towards a negative charge and away from a positive charge. The strength of an electric field is represented by the closeness of the drawn lines. (It can be measured in either volts per metre (Vm^{-1}) or newtons per coulomb (NC^{-1}).

electric storm an atmospheric disturbance in which the air is charged with static electricity which, with clouds, produces thunderstorms.

electrocardiograph see ECG.

electrode a conductor which enables an electric current to be passed into or out of a liquid (as in ELECTROLYSIS) or a gas (gas tube) or a vacuum (valve). See also CATHODE, ANODE and CONDUCTION.

electrodeposition deposition by ELECTROLYSIS of a substance on an ELECTRODE. It is used particularly to deposit one metal onto another as in electroplating where the cathode is the object to be coated, or in electroforming (the production of metal items by metal deposition upon an electrode by means of electrolysis).

electrolysis chemical decomposition achieved by passing an electric CURRENT through a substance in solution or molten form. IONS are created, which move to electrodes of opposite charge where they are freed or take part in a reaction.

electrolyte a compound that dissolves in water to produce a solution that can conduct an electrical charge. The solution becomes able to conduct electricity because the compound undergoes ionization, that is it forms mobile ions. If the substance is completely ionized, it is termed a strong electrolyte, e.g., any strong acid such as sulphuric acid. But if it is only partially ionized it is termed a weak electrolyte, e.g., any weak acid such as ethanoic (acetic) acid.

electromagnet is created when an electric current flows through a coil surrounding a soft iron core, and the latter becomes a magnet. It is therefore not a permanent magnet. Electromagnets are used widely in televisions, loudspeakers and large versions on cranes are used to move and separate scrap metal.

electromagnetic induction the production of an ELECTRIC CURRENT when a conductor is moved through a MAGNETIC FIELD. A current is induced to flow only while there is a changing magnetic field due to the movement of the conductor or magnet. The size of an induced current is dependent on the strength of the magnetic field, the cross-sectional area of the conductor, and both the speed and direction of the relative motion between the conductor and the magnetic field.

electromagnetic waves the effects of oscillating electric and magnetic fields that are capable of travelling across space, i.e. they do not require a medium through which to be transmitted. The spectrum of electromagnetic waves is divided into the following categories:

	wavelength (m)	frequency (Hz)
gamma rays	10^{-10}-10^{-12}	10^{21}- ......
X-rays	10^{-12}-10^{-9}	10^{21}-10^{17}
ultraviolet radiation	10^{-10}-10^{-7}	10^{18}-10^{15}
visible light	10^{-7}-10^{-6}	10^{15}-10^{14}
infrared radiation	10^{-6}-10^{-2}	10^{14}-10^{11}
microwaves	10^{-3}-10	10^{11}-10^{7}
radiowaves	10 - 10^{6}	10^{7}-10^{2}

All electromagnetic waves travel through free space at a speed of approximately 3×10^8 ms^{-1}, known as the SPEED OF LIGHT. The only electromagnetic waves that are readily detected by the eye are visible light waves. These consist of various wavelengths, which correspond to the colours red, orange, yellow, green, blue, indigo and violet. The colour red has the longest wavelength, lowest frequency, and the colour violet has the shortest wavelength, highest frequency. An object that appears to be red in colour, for example, has absorbed all the light waves from the blue end of the spectrum while reflecting the ones from the red end.

electron an indivisible particle that is negatively charged and free to orbit the positively charged NUCLEUS of every ATOM. In the traditional model, electrons move around the nucleus in increasingly large shells. However, current thinking, based on quantum mechanics, regards the electron as moving around the nucleus in clouds that can assume various shapes, such as a dumb-bell (two electrons moving) or clover leaf (four moving electrons). The shape and density of the outermost electronic shell will help determine what reactions are possible between particular atoms and molecules, e.g., whether an atom will easily gain or lose electrons to form an ion.

electronegativity (*plural* **electronegativities**) a measure of the power of an atom within a molecule to attract electrons. Every element in the PERIODIC TABLE is given an electronegativity rating based on a scale in which fluorine is given the highest rating of 4, as it has the most electronegative atoms. Electronegativity differences between the different atoms within a molecule can be used to estimate the nature of the bonds formed between those atoms, i.e., whether it is a COVALENT, IONIC or POLAR COVALENT bond.

electron microscope a microscope which uses a beam of electrons rather than a beam of light striking an object. This gives much greater resolution. In the *transmission* electron microscope the electron beam passes through a very thin slice of the object and an image is created due to the scattering of the beam which is focused and enlarged dramatically onto a fluorescent screen. *Scanning* electron microscopy involves scanning the surface of a sample and the image is generated by secondary electrons. The magnification of the object is less with scanning electron microscopy but a three-dimensional image is created.

electron pair two electrons, one each from the outer shells of two atoms, which are shared by the adjacent nuclei to form a bond.

electrophoresis a method for separating the molecules within a solution using an electric field. Molecules move in an electric field at a speed determined by the ratio of their charge to their mass. All electrophoretic separations involve placing the mixture to be separated onto something porous such as filter paper. An electric field is applied, causing the molecules, now dissolved in a conducting solution, to move at different rates. The various molecules can be identified by comparing their final position on the supporting medium with the position of known standards. Gel electrophoresis uses a synthetic polymer as the supporting medium, and it can be used to separate the different lengths of DNA chains found within the nucleus of a cell. This makes gel electrophoresis an essential technique in "genetic fingerprinting," as the separated DNA strands will contain genes unique to a particular person, thus helping to identify the person from whom the cell came.

electrostatics the part of the science of electricity dealing with the phenomena associated with electrical charges at rest. Any material containing an electric charge that is unable to move from atom to atom is called an INSULATOR (as opposed to a CONDUCTOR, which allows electric charge to flow throughout it). It is possible to negatively charge an insulator such as polythene by transferring electrons to it using a woollen cloth. A similar effect is sometimes produced when a plastic comb is used to comb dry hair. Any item which becomes charged can then induce an opposite charge on, for example, a piece of paper, which will then be attracted to the insulator.

element a pure substance that cannot be separated into simpler, different substances when subjected to ordinary chemical reactions. There are 112 elements known to us, but only 92 of these occur naturally, and the others have been created in laboratories. The elements are classified into the PERIODIC TABLE, according to the number of protons in their nucleus, i.e., their atomic number.

Each element of the periodic table consists of atoms with a unique number of protons in their nuclei.

elementary particle (or **fundamental particle**) a particle which is the basic building block for all matter. To explain nuclear interactions fully, theories have been put forward suggesting the existence of particles in addition to the NEUTRON, PROTON and ELECTRON. Two types of particles are thought to exist, called LEPTONS (which include electrons) and HADRONS (which include protons and neutrons). A further suggestion is that hadrons are made up of the peculiarly named quarks. Quarks are described with even more peculiar terms such as flavour and charm. Although the theory is generally accepted by physicists, quarks have not been confirmed experimentally.

embryo the developmental stage of animals and plants that immediately follows fertilization of the egg cell (ovum) until the young hatches or is born. In animals, the embryo either exists in an egg outside the body of the mother or, as in mammals, is fed and protected within the uterus of the mother.

empirical formula a chemical formula of a compound that shows the simplest ratio of atoms present in the compound. For example, the molecule butane (fourth member of the ALKANE family) has the empirical formula C_2H_5 although its true molecular formula is C_4H_{10}.

emulsion in chemistry, a colloidal solution of one liquid in another in which particles of one are evenly dispersed throughout the other. Emulsions do not retain this arrangement for they separate out if left for some time. Emulsions are found in foods, medicines, paints etc.

endocrine system the network of glands that release signalling substances (HORMONES) directly into the bloodstream. By this method the hormone travels to, and thus affects, distant target cells, as opposed to just affecting cells that surround the active gland. The pituitary gland (at the back of the head, base of the brain), the thyroid gland (in the neck) and the adrenal glands (above both kidneys) are three of the major endocrine glands within the human body. Each secretes different hormones, which will subsequently affect different parts of the body. For example, the pituitary gland secretes growth hormone, the thyroid gland secretes thyroxine, and the adrenal glands secrete the stress hormone called glucocorticoids, all of which control various functions in the body.

endothermic reaction a chemical reaction in which heat energy is absorbed. The required heat energy is supplied by the environment surrounding the reaction, and the products of an endothermic reaction will have stronger bond energies than the original reactants.

energy is the capacity to do WORK. There are various forms of energy, including light, heat, sound, mechanical, electrical, kinetic and potential, but all are expressed in the same unit of measurement, called the JOULE (J). Energy has the capacity to change from one form to another (*energy transfer*), but the original input of energy tends to be greater than the final output during energy transfers. As the law of conservation of energy states that it is impossible to make or destroy energy, the difference in the input/output energy levels is a result of the conversion of some of the input energy into an unwanted form, e.g., heat instead of mechanical energy. The energy content of a system or object can be regarded as the "work done" by it and can be calculated using the following equation:

Work Done (*w*) = Force (*F*) x Distance Moved (*S*)

entropy a measure of the randomness or disorder of a system. It is a natural tendency of the whole universe that allows all energy to be distributed. The entropy of any system will therefore tend to increase as energy is lost at all stages of change. Entropy is illustrated by a hot object in a cold surrounding, where heat flows out of the hot object, rather than the other way around. The greater the disorder, the higher the value for entropy, but at absolute zero entropy is also zero.

enzyme any PROTEIN molecule that acts as a natural CATALYST and is found in the bodies of all bacteria, plants and animals. Enzymes are essential for life as they allow the complex chemical reactions of biochemical processes to occur at the relatively low temperature of the body. Enzymes are highly specific in that they will only act on certain materials (SUBSTRATES) at a particular pH and temperature. For example, the digestive enzymes called amylase (in saliva—turns starch into sugar), lipase (breaks down fats) and trypsin (from the pancreas, helps break down protein) will only work in alkaline conditions (pH > 7), whereas the digestive enzyme, pepsin (in the stomach; breaks down protein) will only work in acidic conditions (pH < 7). Enzymes possess characteristic molecular shapes which allow molecules to fit onto the enzyme. The new molecules become joined and then detach from the enzyme, as a new structure.

eon the largest geological unit of time (see APPENDIX 5) e.g., Phanerozoic which includes the Palaeozoic, Mesozoic and Cenozoic ERAS.

ephemeris a published table providing the projected position and movements of planets and comets, of use to a navigator or astronomer.

epicentre the point or line on the Earth's surface which is directly above the focus of an EARTHQUAKE.

epilepsy a seizure disorder caused by damaged tissue in the brain. The symptoms are in the form of attacks, known as fits, which can include a feeling of numbness, muscular convulsions, inability to speak, etc. Epilepsy can be controlled by certain drugs, but in bygone days, surgery was performed on patients who suffered frequent and extreme attacks in an attempt to control these.

epoch an interval of geological time (see APPENDIX 5) subsidiary to the period, and forming several ages, e.g. the Pleistocene.

equatorial an astronomical telescope which is positioned so that when it is set on a star, that body will be kept in the field of view. This is achieved by setting the telescope to revolve about an axis parallel to the Earth's axis.

equilibrium in a chemical reaction, equilibrium is a condition in which the proportion of the chemicals reacting together and the products being formed is constant, because the rate or speed of the forward reaction is the same as the rate of the reverse reaction. Equilibrium is affected by changes in pressure, temperature and concentration, but is not affected by the addition of a CATALYST (this affects only the speed of the reaction). Equilibrium is also seen elsewhere. For example in physics, equilibrium is when forces acting on a system are equal and the system does not change.

era a unit of geological time which comprises several PERIODS, for example, the Palaeozoic (divided into upper and lower) contains six periods from the Cambrian to the Permian.

erosion the destructive processes which, with WEATHERING, constitute DENUDATION. Specifically, erosion involves the further breakdown and transport of material by water, ice and wind and because of the transportation of material, the continued wearing down of the land surface. The transporting agents thus erode by means of wind laden with sand scouring rock, glaciers containing rocks and boulders grinding down the rocks over which they pass, and rivers excavating their own courses due to movement of rocks, pebbles and particles in the water. Rivers also carry materials in solution.

erythrocyte the term for a red blood. In vertebrates, red blood cells are made primarily in the bone marrow. The erythrocyte differs from other cells in the human body in that just before it is released into the bloodstream, it sheds its nucleus. Erythrocytes contain HAEMOGLOBIN, the protein molecule essential for transportation of oxygen from the lungs to all tissues in the body. Someone who does not have enough erythrocytes circulating in their blood is said to be anaemic. Anaemia can be a result of a defect in the structure of the erythrocyte membrane or lack of iron to form the haem group of the haemoglobin molecule.

escape velocity the velocity required of an object e.g., rocket, space probe, to escape from the gravitational pull of a larger body, e.g., a planet. The necessary velocity depends upon the planet's mass and diameter. For the EARTH it is 11200 ms^{-1} (or 25000 miles per hour) and for the Moon it is much less, 2400 ms^{-1}. The behaviour of a BLACK HOLE is explained by its escape velocity being greater than the SPEED OF LIGHT, so that light cannot escape from it.

escarpment (or scarp) a cliff or steep slope at the edge of an essentially flat or gently sloping area, caused by a combination of original geology and the angle at which the rocks are lying, and subsequent erosion.

esker a steep-sided ridge made of sands and gravels which is the remains of a stream which ran beneath or within a glacier.

ester an organic hydrocarbon compound formed from organic acids. Many esters have a fruity smell and are used for flavourings. Esters are common in nature, as animal fats and vegetable oils are formed from mixtures of esters.

estuary a partially-enclosed stretch of water which is subjected to marine tides and fresh water draining from the land. An estuary is usually created as a DROWNED VALLEY due to a rise in sea level after a period of glaciation. A large amount of sediment is deposited in estuaries and the tidal currents may produce channels, sandbanks and sand waves. Estuaries are rich in all sorts of animal and plant life but they are also ideal sites for commercial development and are often subject to pollution.

ethane the second member of the series called ALKANES. It is an insoluble colourless gas with the chemical formula CH_3CH_3.

ethanol (C_2H_5OH) an alcohol which has a functional hydroxyl group (OH) in place of one hydrogen atom in the structure of ethane, i.e. CH_3CH_2OH. Ethanol is obtained either by fermentation of carbohydrates to form alcoholic beverages, or is prepared commercially from ETHENE (C_2H_4) by adding water and sulphuric acid. Ethanol is also used in the manufacture of foodstuffs and solvents.

ethene the first member of the alkene family, which is an insoluble gas at room temperature. Ethene ($C2H4$) is an important starting chemical in the industrial manufacture of the plastic POLYMER called polythene, i.e. polyethene. It is also used to make many other industrial substances.

ethylene glycol (*also called* dihydroxyethane $HOCH_2.CH_2.OH$) a colourless liquid used in anti-freeze and as coolants for engines. It is used for the manufacture of polyester fibres such as Terylene and elsewhere in the textile industry, and also in printing inks and foodstuffs.

ethyne the first member of the ALKYNE family, it has the chemical formula C_2H_2. It is also known as acetylene and is a highly flammable gas which, when burned with oxygen, will produce the high temperature flame (> 2500°C) characteristic of the oxyacetylene torch needed to cut and weld metals.

eucaryote any member of a class of living organisms (except viruses) that has in each of its cells a nucleus within a membrane. All plants and animals are eucaryotes unlike bacteria and cyanobacteria which are PROCARYOTES.

eugenics the study of how the inherited characteristics of a human population can be improved by genetics, i.e. controlled breeding.

evaporation the process by which a substance changes from a liquid to a vapour. Evaporation occurs when a liquid is heated and some molecules near the surface of it eventually have enough KINETIC ENERGY to overcome the attractive forces of the remaining molecules and escape into the surrounding atmosphere. During evaporation from an open container, the temperature of the liquid falls until heat from the surroundings flows in to replace this heat loss. This explains why swimmers can feel chilled when they emerge from the water; heat energy from the skin is converted to kinetic energy, allowing some water molecules to evaporate. The opposite situation is that heat will speed up evaporation so a puddle soon dries up on a sunny day. Evaporation is an important part of the movement of water on Earth because water evaporates from oceans, rivers, etc. before condensing as clouds.

evaporite sedimentary rocks formed by precipitation from solution during evaporation of lagoons, salt pans (a basin-like area in a semi-arid region) and salt lakes. The least soluble salts such as calcium carbonate ($CaCO_3$) and magnesium carbonate ($MgCO_3$) precipitate first, followed, for example, by (in increasing solubility) sodium sulphate (Na_2SO_4) and then potassium chloride (KCl) and magnesium sulphate ($MgSO_4$). The same principle applies to evaporating sea water when calcium sulphate (or gypsum) is the first to precipitate, followed by ANHYDRITE and HALITE (NaCl). The rock types formed, in addition to gypsum and anhydrite include limestone, dolomite and rock salt (halite).

evening star in general terminology, the name given to Venus (or Mercury) which is seen in the western sky around sunset.

event horizon the boundary of a BLACK HOLE which is thought to be spherical. This marks the line beyond which light cannot escape (see ESCAPE VELOCITY).

evolution the process by which an organism changes and thus gains characteristics that are distinct from existent relatives. Any species of organism will only evolve if:

(a) There has been genetic mutation allowing variation in the genetic information the parent passes on to its descendants.

(b) An individual proves to be more suitable to a particular environment than its relatives, allowing it to survive and reproduce whereas its relatives will become extinct, i.e., NATURAL SELECTION.

exothermic reaction a chemical reaction in which heat energy is released to the surrounding environment. The products of an exothermic reaction will have weaker bond energies and therefore be more stable than the bonds within the molecules of the original reactants.

exponent a symbol, usually a number, that appears as a superscript to the right of a mathematical expression and indicates the power to which the expression has to be raised. For example, the expressions a^5 and 3^4 have the exponents 5 and 4 respectively, and 3^4 is the same as $3 \times 3 \times 3 \times 3$, which is 81.

extrapolate to predict the unknown value of a measurement or function using known values. For example, on a graph, extrapolation involves extending the curve that has been plotted from existing readings beyond the set of known values. The extension uses the same trend to gain an idea of possible values where no readings are actually available.

extrusive rocks a general term that includes rocks of volcanic origin which flow onto the Earth's surface.

F

factor one of two or more quantities that produce a given quantity when multiplied together. For example, the factors of the number 8 are 1, 2, 4, 8.

factorial the multiplication of a series of consecutive numbers from 1 to n inclusive, where n is a whole number. Thus, factorial 5, written as 5! = 1 × 2 × 3 × 4 × 5 = 120. Factorials of much larger numbers are not usually defined.

Fahrenheit (F) a temperature scale, devised by the German physicist Gabriel Fahrenheit (1686-1736), that set the freezing point of water at 32° and the boiling point at 212°. Fahrenheit temperatures can be converted to CELSIUS by the equation $F = 1.8C + 32$.

fall-out radioactive material deposited on the ground from the atmosphere. The source of the radioactive substances is nuclear explosions or escape from nuclear reactors whether through failure to adequately filter coolants or an accident. Radioactive particles settle on material in the air and, in the case of a nuclear explosion, large quantities of dust, soil etc. are sucked up into the central plume of fire and smoke to be deposited hundreds of kilometres from the source, having first risen high in the atmosphere.

farad the unit in which CAPACITANCE is measured. It is usually denoted by the symbol F. For example, the capacitance (C) of a conductor measured in farads is the charge (in coulombs) needed to raise the potential by one volt i.e. 1 farad = 1 coulomb per volt ($1F = 1\ CV^{-1}$). As the farad itself is usually too large a quantity for most applications, the practical unit is the microfarad (μF), or one millionth of a farad.

faraday the quantity of charge carried by one mole of electrons (which is approximately equal to Avogadro's constant x the charge on an electron), which has the value 9.6487×10^4 coulombs. It was named after Michael Faraday (1791-1867), a British scientist whose contributions to physics and chemistry include ELECTROMAGNETIC INDUCTION, electrolysis and MAGNETIC FIELDS.

fatigue of metals the structural failure of metals due to the repeated application of STRESS, which results in a change to the crystalline nature of the metal.

fats a group of organic compounds that exist naturally and are called lipids. Fats occur widely in plants and animals and serve as long-term energy stores. A fat consists of a GLYCEROL molecule and three FATTY ACID molecules, which are known together as a triglyceride and is formed during a condensation reaction (where water is released). Fats are important as energy-storing molecules since they have twice the calorific value of carbohydrates. In addition, they insulate the body against heat loss and provide it with cushioning, which helps protect against damage. In mammals, a layer of fat is deposited beneath the skin (subcutaneous fat) and deep within the tissues (adipose tissue) and is solid at body temperature due to the high degree of saturation. In plants and fish, the fatty acids are generally less saturated and as such tend to have a liquid-like consistency, i.e. oils, at room temperature. A diet that contains a lot of saturated fats is one factor that contributes to heart disease.

fatty acids a class of organic compounds containing a long hydrophobic hydrocarbon chain (one which is not soluble in water) and a terminal carboxylic acid group (COOH) which is extremely hydrophilic (water soluble). The chain length ranges from one carbon atom (HCOOH; methanoic acid), to nearly thirty carbon atoms, and the chains may be SATURATED or UNSATURATED. As chain length increases, melting points are raised and water solubility decreases. However, both unsaturation and chain branching tend to lower melting points. Fatty acids have three major functions in the body:

(1) They are the building blocks of PHOSPHOLIPIDS (lipids containing phosphate) and glycolipids (lipids containing carbohydrate). These molecules are important components of biological membranes, creating a lipid bilayer which is the structural basis of all cell membranes. The (hydrophilic) 'heads' of the molecules are in contact with water while the (hydrophobic) 'tails' point into the membrane.

(2) Fatty acid derivatives (that is, compounds made from them) serve as hormones and messengers within cells.

(3) Fatty acids serve as fuel molecules. They are stored in the CYTOPLASM of many cells in the form of triglycerides (three fatty acid molecules joined to a glycerol molecule) and are broken down, as required, in various energy-yielding reactions, which produce energy.

fault a geological feature which is essentially a flat plane but which may take up any position. It is caused by the brittle deformation of rocks which are displaced on one side of the fault plane relative to the other. The displacement is usually parallel to the fracture plane. Movement along faults may vary from millimetres to kilometres. The measurement of the movement is often made by referring to the horizontal and vertical components of movement. There are several types of fault, depending upon the sense of movement.

fault breccia a zone composed of broken, angular and crushed rock fragments generated by movement along a fault. Fault breccia is loosely held together (or incohesive) but, at depth, the fault breccia may become lithified (that is, turned into a rock) due to higher pressure and temperatures and, with some recrystallization, will form a crush breccia.

feedback mechanism a control mechanism that uses the products of a process to regulate that same process by activating or repressing it. Almost all homeostatic mechanisms (see HOMEOSTASIS) in animals operate by negative feedback. This is when a variation of certain conditions from the normal state triggers a response that tends to oppose the variation, bringing the system back to normal. For example, it operates during the release of hormones to maintain steady blood sugar levels. Another example is sweating which is 'switched on' by the brain when it registers that the body is too warm. Positive feedback is found less often as a biological control mechanism. Here, a variation from the normal causes that variation to be amplified, and this is usually a sign that the normal control mechanisms have broken down.

feldspar a very important group of rock forming minerals which are aluminium silicates with combinations of calcium and sodium (Ca/Na) or sodium and potassium (Na/K). They occur in many rock types but are essential components of igneous rocks and in certain types of granite are seen as large pink or white crystals.

fermentation a form of ANAEROBIC RESPIRATION, which converts organic substances into simpler molecules, generating energy in the process. Fermentation, carried out by certain organisms such as bacteria and YEASTS, is the conversion of sugars to alcohol in the process known as alcoholic fermentation and carbon dioxide is released as a by-product. Lactic acid fermentation occurs in the muscles of higher animals when the oxygen requirement exceeds the supply and sugar is converted into lactic acid. In industry, fermentation is important in baking and in beer and wine production, and these use large quantities of yeast. Fermentation is also used in biotechnology and in the manufacture of bread and yoghurt.

Fermi, Enrico (1901-1954) an Italian-born American physicist who was awarded the Nobel Prize in 1934 for his discovery that stable elements would become unstable when bombarded with NEUTRONS as they have become radioactive. His later research contributed to the harnessing of atomic energy and to the construction of the atomic bomb.

fertilization is the fusion or joining together of the male (sperm) and female (egg or ovum) sex cells which is the essential part of SEXUAL REPRODUCTION. Fertilization describes the process in which the two cells come together to become one and it sets in motion a chain of events, (involving further cell division and growth) which eventually gives rise to a new individual. It is a common event in both plants and animals and enables genetic "mixing" to occur as the new organism receives its characteristics from each parent. In many animals, e.g., most fish, fertilization is described as *external* as the eggs are laid outside the body and sperm are shed over them. In many other animals, e.g. MAMMALS, fertilization is *internal* as the male sex cells are released inside the body of the female.

fertilizers are chemicals added to the soil to improve crops and their yield and the growth of plants and flowers. Fertilizers replace the nutrients in the soil that are extracted by growing plants. Modern farming is very intensive and natural processes are unable to provide all the necessary nutrients required. In addition to carbon, hydrogen, oxygen there are other *essential elements* such as nitrogen, phosphorus, potassium, calcium, magnesium and sulphur; plants require *trace elements* (perhaps in parts per million quantities) such as iron, boron, manganese, zinc, copper, molybdenum and chlorine. Artificial fertilizers make up a lot of the deficits in these elements.

Ammonium sulphate is an important nitrogenous fertilizer (i.e., nitrogen supplying) and other chemicals used include sodium nitrate, urea and ammonia. *Superphosphates* contain phosphorus, the chemical used being calcium hydrogen phosphate, $Ca(H_2PO_4)_2$.

It is essential that fertilizers are used correctly and that they are not overused as the excess nutrients can have detrimental effects on the land, and particularly on streams and rivers. As the nutrients drain into water they encourage the growth of algae and surface plants which choke the stream and results in a lack of oxygen in the water which eventually kills animal and plant life beneath the surface.

fibrinogen a PROTEIN in the blood, which causes it to form clots, due to action by the ENZYME thrombin.

field see **electric field, magnetic field**.

filter paper a pure cellulose paper used in the laboratory for the separation of solids from liquids by filtration.

filtrate the liquid remaining after filtration, having been separated from a solid/liquid mixture. The solid material remaining is called the residue.

filtration separation of a solid from a liquid by passing the mixture through a suitable separation medium e.g. FILTER PAPER which holds back the solid and permits the liquid to pass through. A gas may also be filtered. Filtration processes are used in the home and in numerous industrial processes including water purification.

finder a small telescope used in conjunction with, and fixed to, a large telescope, for finding the required object and fixing it in the centre of the viewing field of the large telescope.

fissile a fissile element is one which undergoes NUCLEAR FISSION.

fission the spontaneous or induced splitting of a heavy nucleus (such as uranium) into two fragments during a nuclear reaction, which subsequently releases vast quantities of energy. Nuclear fission is induced (encouraged to happen) by irradiating nuclear fuels like uranium with NEUTRONS in a device called a nuclear reactor, and this process is accompanied by the emission of several neutrons. These neutrons in turn cause fission of another nucleus, which, under suitable conditions, can result in a CHAIN REACTION. For a chain reaction to occur, there has to be more than a *critical mass* of uranium, otherwise too many neutrons escape without causing further fission. The energy released is in the form of heat, and it can be harnessed and used to produce electricity by making steam, which is used to drive turbines.

flagella (*singular* **flagellum**) long thread-like extensions from the surface of a cell, similar to cilia. Protozoa (single-celled animals) use flagella for locomotion.

flame test a simple test used in chemistry for the detection of metals which is useful for distinguishing between different metals. A small quantity of the unknown sample is placed on a platinum wire and held in a flame, and the resultant colour is characteristic of the particular metal. For example, when sodium compounds are held into a flame, the flame burns with a bright yellow colour. Potassium gives a violet flame, and lithium and strontium give a red flame. Although lithium and strontium appear similar, the light from each can be resolved (separated) into different colours by using a prism, and this resolution easily distinguishes the two elements.

flare a sudden outburst of radiation from a star, or more particularly from the lower atmosphere of the SUN.

flash distillation a technique which involves spraying a liquid mixture into a heated chamber at low pressure. This enables rapid removal of solvent because it is more volatile. The technique is used in the petroleum industry.

Fleming, Sir Alexander (1881-1955) a Scottish bacteriologist who discovered the antibiotic penicillin in 1928, for which he was awarded the Nobel Prize in 1945.

flint a form of SILICA (silicon dioxide, SiO_2) which is composed of minute crystals occurring commonly as nodules or bands in CHALK deposits. It breaks with a conchoidal (curved) fracture to produce sharp edges, a property which was utilised by prehistoric man in the making of tools and weapons.

flood plain the flat or almost flat surface at the bottom of a river valley within which the river flows. It is the area that becomes covered when the river floods. It is formed by the progressive development of the river as it meanders laterally (moves from side to side) and is made up of sediment carried by the river which often show a fining-upwards sequence from the underlying rock, through coarse gravels, then sand followed by clays and silts. This rhythmic sequence of sediments may be repeated.

fluid any substance that flows easily and alters its shape in response to outside forces. All gases and liquids are fluids. In liquids, the particles move freely but are restricted to the one mass, which occupies almost the same volume. In gases, however, the particles tend to expand to the limits of their containing space and thus do not keep the same volume. Because of those properties fluids are useful in driving machines.

fluorescence the property of certain substances that absorb radiation (e.g. ultraviolet) and then emit radiation as visible light. This occurs because molecules become excited through absorbing the incoming light of a particular WAVELENGTH, and their energy level is raised. When the energy level falls, the radiation is emitted at a different, usually greater, wavelength.

fluorocarbons (see *also* CFC) a group of organic compounds in which fluorine replaces some or all of the hydrogen atoms in the hydrocarbon. The resulting compounds are similar to the original hydrocarbon in some ways, but are unreactive and thermally stable. They are used in oils and greases for applications where ordinary materials would be attacked. There are fluorine-containing polymers, and chlorofluorocarbons have been used as refrigerants and aerosol propellants, although these latter uses are being discontinued due to the effect of the CFCs on the OZONE LAYER.

fluviatile deposits sediments deposited by a river, often comprising sands and gravels.

flux a substance added to a solid to assist in its fusion, e.g. cryolite (naturally occurring sodium aluminium fluoride) is added to bauxite from which ALUMINIUM is extracted.

focal plane in an optical system the plane in which the image of an object is formed and therefore the position for the film or plate. The focal plane is at right angles to the principal axis of a lens (system) which runs through the centre of the lens and at 90° to the long dimension of the lens.

focal point the point at which converging rays meet on the axis of a lens system.

focus the point at which rays converge after passing through an optical system.

foetus the term used to describe the developing young in the mammalian uterus, from the post-embryonic period until birth. In humans, this period is from about 7-8 weeks after FERTILIZATION.

fog a suspension of water droplets (up to 20µm diameter) in the atmosphere when the air is near to saturation, causing visibility to fall below 1 kilometre. Fog formation is enhanced by the presence of smoke particles which act as nuclei for the condensation of the droplets. The condensation may be caused by cooling of the ground or warm air moving over cold ground or water.

föhn wind a general term for a warm, dry wind in the lee of a mountain range.

fold in geology a fold is produced when originally flat beds of rock become curved or bent due to the application of immense forces over a long period of time. There are many types of fold, depending on different factors such as the type of rock, the intensity of the deforming forces etc., but there is a basic set of descriptions which can be applied to all folds. The zone of maximum curvature is the hinge and the limbs lie between hinges or (for a single fold) on either side of the hinge. Thus in a series of folds, each limb is common to two adjacent folds. Two other useful features are the axial plane and axis. The axial plane for one fold is an imaginary feature which bisects the angle between both limbs and is the same distance from each limb. The fold axis is the line created by the axial plane intersecting the hinge zone. Folds range in size from a few millimetres to many hundreds of metres.

folic acid a compound which forms part of the VITAMIN B complex. It is involved in the formation of some AMINO ACIDS and nucleic acids in the body and is used in the treatment of anaemia. In the diet it is found in green vegetables, whole wheat products, peas and beans.

food chain in simple terms, a food chain is the route by which energy is transferred through a number of organisms by one eating another from a lower level (called a *trophic* level). At the base of the chain are the primary producers which are the green plants. These are able to use energy from the sun to manufacture food substances from carbon dioxide and water. These food substances (glucose, cellulose and starch) are made use of by animals at the next level in the chain. The herbivores which eat the green plants are known as primary consumers. These in turn are eaten by carnivores (flesh-eating animals) which are called secondary consumers. There may be more than one level of secondary consumer (a flesh-eater may itself be eaten by a larger carnivore) ending up with animals at the end of the chain, e.g. lions, which are not preyed upon and are called the top predators. However, when these and all organisms die, they are eaten by scavenging animals and the remains eventually broken down by micro-organisms so none of the energy is lost but is used again. Food chains are often highly complicated and all of those that exist in a given environment are inter-linked and form a food web. A food chain should be perfectly balanced with many more organisms at the lower levels than at the higher ones. Sometimes the natural balances are upset and this may be due to human interference. For example, in Britain there are no large carnivores such as wolves because they were hunted and killed off during the Middle Ages. In Scotland red deer numbers are too high and (although the situation is complicated) these would, at one time, have been hunted by wolves.

force the push or pull exerted on a body in a particular direction. The force may alter the state of motion by causing the velocity of the body to increase or decrease (it may accelerate or decelerate). An object will continue to move at a constant speed in a straight line (excluding friction) unless another force acts upon it. The unit of force is the newton, given by $F = ma$ where m is the MASS of the body and a is its ACCELERATION.

formula (plural **formulae**) a law or fact used in science and mathematics, denoted by certain symbols or figures. In mathematics and physics, it is expressed in algebraic or symbolic form. In chemistry, there are three principal types of formulae—EMPIRICAL FORMULA, MOLECULAR FORMULA, and STRUCTURAL FORMULA; it provides a sort of shorthand for writing down the constituents of a compound.

fossil the remains of once-living plants and animals, or evidence of their existence, preserved in the rocks of the earth's crust. Palaeontology is the name given to the study of fossils and it has proved useful in the study of evolutionary relationships between organisms, and in the dating of geological strata. There is a vast range of fossils from plants to the shells of marine organisms and the bones of huge dinosaurs.

fossil fuels these are NATURAL GAS, PETROLEUM (oil) and coal and they form the major fuel sources today. They are formed from the bodies of aquatic organisms that were buried and compressed on the bottoms of seas and swamps millions of years ago. Over time, bacterial decay and pressure converted this organic matter into fuel.

Hard coal, which is estimated to contain over 80 per cent carbon, is the oldest variety and was laid down up to 250 million years ago. Another, younger variety (bituminous coal) is estimated to contain between 45 per cent and 65 per cent carbon. The fuel values of coal are rated according to the energy liberated on combustion. Coal deposits occur in all the world's major continents, and some of the leading producer countries are the United States, China, Russia, Poland and the United Kingdom.

Natural gas consists of a mixture of HYDROCARBONS, including METHANE (85 per cent), ETHANE (about 10 per cent) and PROPANE (about 3 per cent). However, other compounds and elements may also be present, such as carbon dioxide, hydrogen sulphide, nitrogen and oxygen. Very often, natural gas is found in association with petroleum deposits. Natural gas occurs on every continent, the major reserves being found in Russia, the United States, Algeria, Canada and in counties of the Middle East.

Petroleum is an oil consisting of a mixture of HYDROCARBONS and other elements (e.g. sulphur and nitrogen). It is called crude oil before it is refined. This is done by a process called FRACTIONAL DISTILLATION, which produces four major fractions:

(1) Refinery gas, which is used both as a fuel and for making other chemicals.

(2) Gasoline, which is used for motor fuels and for making chemicals.

(3) Kerosine (paraffin oil), which is used for jet aircraft, for domestic heating and can be further refined to produce motor fuels.

(4) Diesel oil (gas oil), which is used to fuel diesel engines.

The known residues of petroleum of commercial importance are found in Saudi Arabia, Russia, China, Kuwait, Iran, Iraq, Mexico, the United States, and a few other countries.

Together, the fossil fuels account for nearly 90 per cent of the energy consumed in the United States. As coal supplies are present in abundance compared with natural gas or petroleum, much research has gone into developing commercial methods for the production of liquid and gaseous fuels from coal.

fossilization is the formation of a FOSSIL. Organisms tend to undergo some changes after death and are not usually preserved whole. In particular the soft parts will decay and skeletal parts are often changed. Sediment may flatten an organism and porous structures may be replaced by minerals. Recrystallization commonly occurs, replacing the fine structure of shells, and the mineral aragonite (a form of CALCIUM CARBONATE) often changes to the more stable CALCITE. Skeletal parts may leave impressions or moulds in sediments and these may be internal or external. In addition, burrows, trails and similar evidence of organisms can be preserved as TRACE FOSSILS.

fraction a quantity that is only part of a whole unit. It is written as x/y where x and y are whole numbers and x is called the numerator and y the denominator (which can never be zero).

fractional distillation (*also called* **fractionation**) is the process used for separating a mixture of liquids into its different parts (called fractions) by distillation. The liquid to be separated is placed in a flask or distillation vessel to which a long vertical column (fractionating column) is attached. The liquid is boiled, causing it to turn into vapour, and as the vapour rises up the column it cools and condenses and runs back into the vessel. The vapour in the vessel continues to rise, and as it does so it passes over the descending liquid. This eventually creates a steady temperature gradient, with temperature decreasing towards the top of the column. The components of the mixture that vaporize easily (and which have low BOILING POINTS) are said to be more volatile and are found towards the top of the column, where the temperature is lowest, while less volatile components are found towards the bottom of the column. At various points on the column, the different fractions can be drawn off and collected. Those components with appreciably different boiling points will be separated into the different fractions.

Petroleum contains a mixture of hydrocarbons, and fractional distillation is used to separate the components into fractions such as gasoline and kerosine. It was fractional distillation of liquid air that led to the discovery of three of the NOBLE GASES (neon, krypton and xenon), and today the process is used to obtain large quantities of molecular oxygen (O_2) required for commercial purposes. Air is first compressed and cooled (which freezes out carbon dioxide and water) and then fractioned in a liquid air machine. Molecular oxygen can be obtained since the other components of air (nitrogen and argon) are more volatile and can be removed from the top of the fractionation column as gases.

fraternal twins twins that are not identical which develop when two ova (eggs) are fertilized at the same time. This occurs when two ova have matured and have been shed simultaneously, and the resultant twins resemble each other only to the same extent as brothers and sisters born at different times.

Fraunhofer lines in the continuous spectrum of light from the Sun, Fraunhofer lines occur as dark lines due to the absorption of certain wavelengths of the light by elements in the chromosphere (the layer of gas around the Sun). The major lines are due to the presence of calcium hydrogen, sodium and magnesium.

free energy (*also called* **Gibb's free energy**) a thermodynamic quantity used in chemistry, which gives a direct measure of spontaneity of reaction in a reversible process. It is defined by the equation $G = H - TS$, where G is the energy liberated or absorbed, H is the enthalpy, S is the ENTROPY, and the system is measured at constant pressure and temperature (T). As a reaction proceeds, reactants form products, and H and S change. These changes, denoted by ΔH and ΔS, result in a change in free energy, ΔG, given by the equation $\Delta G = \Delta H - T\Delta S$. If ΔG is a large negative number, the reaction is spontaneous, and reactants transform almost entirely to products when equilibrium is reached. If ΔG is a large positive number, the reaction is not spontaneous, and reactants do not give significant amounts of products at equilibrium. If ΔG has a small negative or positive value (less than 10 kilojoules), the reaction gives a mixture of both reactants and products in significant amounts at equilibrium.

freeze-drying a process used when dehydrating heat-sensitive substances (such as food and blood plasma) so that they may be preserved without being damaged during the process. The material to be preserved is frozen and placed in a VACUUM. This causes a reduction in pressure, which in turn causes the ice trapped in the material to vaporize, and the water vapour can be removed, producing a dry product. For most solids, the pressure required for vaporization is quite low. However, ice has an appreciable vapour pressure, which is why snow will disappear in winter even though the temperature is too low for it to melt.

frequency (f) the number of complete wavelengths passing any given reference point on the line of zero disturbance in one second. For example, the frequency of an OSCILLATION, such as a wave, is the number of complete cycles produced in one second, the unit of which is the hertz (Hz). The wave equation is given by $c = f\lambda$, where c = the velocity of the wave and λ is its wavelength. For example, if waves have a wavelength of 2 metres and travel with a velocity of 10 metres per second, then the frequency of the wave motion is 5 Hz.

friction the force that opposes motion and always acts parallel to the surface across which the motion is taking place. Unless a force is exerted to keep an object moving, it will tend to slow down due to the opposing force of friction. Friction can therefore be thought of as a negative force, causing negative acceleration. Frictional forces between two solids or between a solid and a liquid are much greater than those between a solid and air. The hovercraft is a vehicle that exploits this fact by travelling on a cushion of air, thus reducing friction. However, friction is useful. It enables us to walk because of the resistance between the ground and our shoes and the friction between the road and the tyres of a car enables the car to move forward, ensuring the wheels do not slip and spin.

friction layer the layer in the atmosphere in which the effects of surface friction are registered. It extends to approximately 2000 feet.

front (or *weather front*) in meteorology, the boundary between large masses of air with different properties, due to differences in temperature and pressure. Fronts can be identified by the air masses separated at the front, their stage of development and the direction of their advance (see COLD FRONT and WARM FRONT).

fuel any material that, when treated in a particular way, releases stored energy in the form of heat. The FOSSIL FUELS and organic fuels (e.g., wood and waste material) produce energy by combustion in the presence of oxygen, releasing carbon dioxide and water. Nuclear fuels, such as uranium and plutonium, release large amounts of heat during nuclear reactions when chemical changes occur within the atom (see FISSION).

function a mathematical term used when there is a link or relationship between two or more VARIABLES. For example, if two variables are x and y, and there is an associated value for y for every value of x, then y is said to be a function of x. The values of x are termed the domain of such a function, and the range of that function is the term given to the corresponding set of y values.

functional group in organic chemistry, an arrangement of atoms joined to a carbon skeleton, which gives an organic compound its particular chemical properties. Compounds with the same functional groups are classed together because of their similar properties. For example, compounds with the functional group NH_2 are classed as amides; those with carbon-carbon double bonds are classed as ALKENES; those with the -OH group are classed as ALCOHOLS.

fundamental particle *see* **elementary particle**.

fungus (*plural* **fungi**) all fungi are simple organisms which may be one cell or exist as threads (or *filaments*) of many cells. They were once classified as simple plants but as they contain no CHLOROPHYLL and cannot PHOTOSYNTHESIZE, they are now placed in their own kingdom—fungi. Fungi absorb their food from other organic material and are vital in the breakdown and recycling of organic substances. They are essential in that they make minerals available to the roots of growing plants. Fungi are vital organisms for man, being used in the processes of BIOTECHNOLOGY and FERMENTATION. Some are harmful causing diseases in plants and animals, others are parasites and a few are edible, e.g., mushrooms. The scientific study of fungi is called *mycology*.

fuse a device for maintaining the CURRENT in a CIRCUIT by preventing it from rising too high if a fault should occur. It is simply a thin metal wire of low MELTING POINT so that the heat generated from too high a current melts the wire, causing the circuit to break and the current to fall to zero. Fuses have different ratings according to the thickness of the wire. The fuse rating is the maximum current that can flow through the fuse without causing the circuit to break.

fusion a nuclear reaction in which unstable nuclei combine to create larger, more stable nuclei with the release of vast amounts of energy. For the reaction to occur, the nuclei have to collide, and this means that the nuclei must have very high KINETIC energies to overcome the repulsive forces between them. NUCLEAR FUSION occurs in the hydrogen bomb (fusion bomb), and at temperatures of about 100 million degrees centigrade, the reaction becomes self-sustaining, that is, it keeps itself going. Our Sun is essentially a nuclear fusion reactor where in the core hydrogen is converted into helium and enormous quantities of energy are released.

G

galactic halo a collection of stars, gas and dust, nearly spherical in shape, which has the same centre as the galaxy. It contains population II stars (see POPULATION TYPES) and creates a lot of the background radio emissions.

galactic noise background noise from galactic sources.

galactic plane the plane which passes through the centre of the MILKY WAY, as far as that can be determined.

galactic rotation the rotation of the GALAXY, and all its gas, stars and dust, which increases towards its centre. The rotation near the SUN is approximately 250 kms^{-1} and to make one galactic orbit takes the Sun in the region of 250 million years.

galaxy (*plural* **galaxies**) specifically, the name given to the band of stars, numbering one hundred thousand million bodies, which includes the Sun, and is alternatively called the Milky Way. The galaxy has a spiral structure and is approximately one hundred thousand LIGHT YEARS across. In general usage it refers to a collection of stars held together by gravitational forces. They form different shapes including spiral, elliptical and irregular galaxies and range from dwarf galaxies with 100 000 stars to massive galaxies such as the giant elliptical M 87 which has three thousand billion stars.

gale a wind of force 8 on the BEAUFORT SCALE (more than 30 knots).

galena the commonest ore of the metal lead, occurring as lead sulphide (PbS) often in the form of grey cubic crystals. It is often found in association with blende (zinc sulphide) and silver sulphide may be present in quantities sufficient to warrant processing for silver.

gall bladder see bile.

galvanic cell an electrochemical cell that generates energy.

galvanizing the industrial process by which one type of metal is coated with a thin layer of another, more reactive metal. Galvanizing is performed for the purpose of protection, as the more reactive metal coating will corrode before the underlying metal. For instance, sheets of iron and steel are often coated with the more reactive metal, zinc. Even when the zinc coating becomes damaged, the underlying iron or steel will be protected. The galvanizing can be done by dipping metal in molten zinc, or by electrolysis in which the part to be galvanized forms the cathode in a solution of zinc ions.

gamete the reproductive cell of an organism. Gametes can be either male or

female, and these specialized cells are HAPLOID in number but unite during fertilization, producing a DIPLOID ZYGOTE that later develops into a new organism. In
higher animals, the male and female cells are called sperm and ova respectively,
whereas in higher plants they are known as pollen grains and egg cells respectively. In some organisms there is essentially one type of gamete that is capable
of developing into a new individual without fertilization. These gametes are usually diploid, as in the case of certain lower plant groups, e.g. many forms of algae.

gamma ray a type of ELECTROMAGNETIC radiation released during the radioactive
decay of certain nuclei and caused by energy changes inside the nuclei of atoms.
The rays released are the most penetrating of all radiations, requiring about
twenty millimetres of lead to halve their intensity. The gamma rays are useful
for sterilizing substances and in the treatment of cancer. They have the shortest
wavelength of any wave in the electromagnetic spectrum, i.e., 10^{-10} to 10^{-12} metres, and are in effect high energy photons.

ganglion (*plural* **ganglia**) a mass of nervous tissue which contains nerve cell
bodies (the part of the nerve cell with the nucleus) and SYNAPSES. Ganglia form
part of the central nervous system (CNS) in invertebrates, occurring along the
nerve cords but in vertebrates they occur outside the CNS in the main. Ganglia
enable parts of the nervous system to control certain activities without needing the whole system.

garnet (the) garnet (family) is a widely-occurring group of minerals which are
found particularly in METAMORPHIC rocks but also in igneous rocks and deposits
that have been subject to weathering and erosion, such as beach sands. They
are silicates of calcium, magnesium, iron or manganese with iron, aluminium,
chromium or titanium. Garnets vary enormously in colour depending on composition and include a number of semi-precious varieties: pyrope (red, containing magnesium and aluminium); grossular or cinnamon stone (light cinnamon
containing calcium and aluminium); and almandine (brownish-red, containing
iron and aluminium). Garnets that are not suitable for gems are used for
abrasives.

gas is the fluid state of matter (solid and liquid being the others). Gases are capable of continuing expansion in every direction because the molecules are held
together only very loosely. A gas will therefore fill whatever contains it and because the molecules move around rapidly and at random, they bump into each
other and the walls of the container which results in a PRESSURE being exerted on
the walls. If a certain amount of gas in a container is put into another container

half the size, the pressure doubles (if the temperature is constant). Also, heating a gas in a container increases the pressure. It can be seen therefore that the temperature, pressure and volume of a fixed MASS of gas are all related and many years ago, early experimentation with gases resulted in three GAS LAWS.

gas laws the rules that relate to the pressure, temperature and volume of an ideal gas, allowing useful information about a gas to be gained by calculation instead of by experimentation. The laws are termed BOYLE'S LAW, CHARLES' LAW, and the pressure law. The pressure law states that, when a gas is kept in a constant volume, the pressure of that gas will be directly proportional to the temperature. All three laws can be combined in an equation known as the universal gas equation, which allows gases to be compared under different temperatures and pressures, i.e., $pV = nRT$, where p, V and T relate to pressure, volume and temperature respectively, n is the amount of gas under investigation, and R is the universal molar gas constant, which has the value of $8.314 \, JK^{-1}mol^{-1}$.

gastrin a HORMONE secreted in the stomach which stimulates the gastric glands of the stomach to produce gastric juice. The presence of food in the stomach is the initial stimulus, and the gastrin controls the digestive process. Gastric juice is a mixture of hydrochloric acid, certain salts and some ENZYMES e.g. PEPSIN, which catalyse the breakdown of protein.

Geiger tube or **Geiger-Müller tube** an instrument, named after the German physicist Hans Geiger (1882-1945), that can detect and measure radiation. The tube contains an inner electrode and a cylindrical outer electrode filled with a gas at low pressure. Any radiation enters the tube through a mica (a silicate mineral that forms thin sheets) window, causing an electrical pulse to travel between the electrodes. These pulses are detectable when the Geiger tube is connected to an electronic circuit, called a scaler, which records the total radiation in the area in a given time.

gel a jelly-like material resulting from the setting of a COLLOIDAL solution. The VISCOSITY is often such that the solution may have properties more like solids than liquids. Two examples of gels are gelatin and silica gel.

gene one of the chemically complex units of hereditary information, found at a certain location on a CHROMOSOME, that is responsible for the transmission of information from one generation to the next. Each gene contributes to a particular characteristic of the organism, and gene size varies according to the characteristic that it codes for. For example, the gene that makes up the codes for the hormone called insulin, consists of 1700 base pairs on a DNA molecule.

gene cloning a method of GENETIC ENGINEERING in which certain genes are extracted from host DNA and introduced into the cell of another host. All the descendants of the genetically transformed host cell will produce a copy of the gene. The transformed gene is thus said to have been cloned (see also CLONE).

gene flow the transfer of genes between populations via the GAMETES. Gene flow contributes to variation in a population as it can lead to a change in the frequency of particular forms of gene present within that population. This in turn is a factor that contributes towards EVOLUTION, as the characteristics of an organism are affected. Therefore, gene flow can be a benefit, as it can help an organism inherit new characteristics that may be beneficial to its survival.

generator see **dynamo**.

genetic adaptation see **adaptation**.

genetic engineering the branch of biology that involves the artificial modification of an organism's genetic make-up. The term covers a wide range of techniques, including selective plant and animal breeding, but it is especially associated with two particular techniques:

(1) The transfer of DNA from one organism to a different organism in which it would not normally occur. For example, the gene that codes for the human hormone, insulin, has been successfully incorporated into bacterial cells, and the bacteria produce insulin.

(2) Recombination of DNA between different species in the hope of producing an entirely new species. For instance, cells of the potato and tomato plants, which have had their cell walls removed, have been successfully cultured and made to fuse together using a variety of experimental procedures. Such cells can grow successfully and develop into a new species of plant that has been called the pomato. Although crossing the species barrier is an important breakthrough in the field of genetic engineering, there are strict governmental regulations regarding the release of such species into the environment since the consequences cannot be predicted.

genetic recombination the exchange of genetic material during meiosis, with the result that the GAMETES produced have combinations of genes that are not present in either parent. This rearrangement of genes allows for variability in a species, and in each generation an almost infinite variety of new combinations of different genes are created. Such new combinations of genes can confer enormous benefits to an organism when conditions change. For example, only a tiny number of a population of locusts have specific combinations of genes that enable them to survive potent pesticides. When such insects reproduce, they produce resistant populations—a major problem in the world of agriculture.

genome the total genetic information stored in the CHROMOSOMES of an organism. The number of chromosomes is characteristic of the particular species of organism. For instance, a man has 23 pairs of HOMOLOGOUS CHROMOSOMES (containing about 50,000 genes), domestic dogs have 39 pairs, and domestic cats have 19 pairs. In each case, one pair of chromosomes constitute the SEX CHROMOSOMES, and the remaining pairs are the AUTOSOMES.

genotype the particular combination of the genes in an individual's genetic make-up.

geocentric a descriptive term meaning any system which has the centre of the Earth as its reference point.

geocentric parallax in astronomy, the movement of an observer due to the Earth's rotation causes an apparent change in the position of a heavenly body. This is the geocentric parallax.

geochemistry a part of GEOLOGY that deals with the chemical make-up of the Earth, including the distribution of elements (and ISOTOPES) and their movement within the various natural systems (ATMOSPHERE, LITHOSPHERE, etc.). Recently, it has also been taken to include other planets and moons within the solar system. The geochemical cycle, in broad terms, illustrates the way in which elements from MAGMA (the 'starting point') move through different processes and geochemical environments.

geochronology the study of time on a geologic scale through the use of *absolute* and *relative age-dating* methods. Relative dating deals with the study and use of FOSSILS and SEDIMENTS to put rock successions/sequence in order. Absolute methods provide an actual age for a rock, using radioactive elements which, by knowing their decay rate (HALF-LIFE), enables the necessary calculations to be made.

geode a rounded 'rock' (similar to a large potato in size and appearance), which on examination proves to be hollow and contains mineral crystals growing from the wall to the centre. Their unimpeded growth into the cavity often produces crystals with perfect HABIT (shape), which are valued by collectors.

geodesic in geometry, the shortest path between two points.

geodesy essentially a combination of mathematics and surveying involving the measurement of the shape of the Earth (or large areas of it).

geography the study of the Earth's surface, including all the land forms, their formation and associated processes, which comprise *physical* geography. Such aspects as climate, topography and oceanography are covered. *Human* geography deals with the social and political aspects of the subject including populations and their distribution. In addition, geography may cover the distribution and exploitation of natural resources, map-making and remote sensing.

geology the scientific study of the Earth, including its origins, structure, processes and composition. It includes a number of topics which have developed into subjects in their own right: geochemistry, mineralogy, petrology (study of rocks), structural geology, geophysics, palaeontology, stratigraphy, economic and physical geology. Charles Lyell (1797–1875) was an influential figure in the early years of geological study and wrote *Principles of Geology*.

geometry a major branch of the mathematical sciences, which involves the study of the relative properties of various shapes. For example, the calculations used to determine the size of the angles and the area of a triangle.

geophysics the study of processes within, and the properties of, the Earth. Included within the discipline are seismology, magnetism, gravity, hydrology, geochronology and studies of heat flow.

geostationary a term referring to an orbit around the Earth, in which a satellite stays in the same position relative to the Earth because it rotates at the same speed as the planet.

geothermal gradient the increase of temperature with depth, into the earth. The gradient varies with location, from continents to oceans and depends greatly upon the earth movements and volcanic activity of a region. In general the gradient is within the range 15° to 40°C per kilometre. This increase in temperature can be put to good use when it is harnessed as geothermal energy. This is when water at depth becomes heated and can either be used directly, or the steam from it can be taken off to generate power.

geotropism a growth movement, exhibited by plants in response to the force exerted by GRAVITY. Plant roots are termed positively geotropic since they grow downwards, whereas plant shoots generally grow upwards (towards sunlight) thus displaying negative geotropism.

germination the start of growth in a dormant structure, e.g. a seed or spore. Various factors can break seed dormancy, such as specific temperatures, exposure to light, or rupture of the seed coat, all of which depend on the species from which the seed is derived.

gestation period the period from conception to birth in mammals, which is characteristic of the species concerned. For instance, dogs have a gestation period that on average is 63 days, whereas that of the blue whale is 11 months.

Larger mammals tend to have longer periods of gestation and the offspring require greater care for a longer time before they can live independently.

geyser a small fissure or opening in the Earth's surface, connected to a hot spring at depth, from which a column of boiling water and steam is ejected periodically. The mechanism of eruption is created by hot rocks heating water to boiling point at the base of the column before the top. Vapour bubbles rise through the column pushing out water and steam at the top, with considerable force. This reduces pressure at the base of the column and boiling continues. Minerals dissolved in the water (calcium carbonate, or silica) are deposited around the mouth of the geyser.

giant star a star in a spectral class (see SPECTRAL TYPES) which is brighter than the main stars in the class.

gibberellin see **hormone**.

Gibb's free energy see **free energy**.

glacial action all processes that are related to the action of a glacier, including accumulation of crushed rock fragments as moraines and the physical actions, e.g., grinding, scouring and polishing which are all due to the incorporation of rocks into the ice, which then act on the rock below and at the sides of the glacier.

glacial deposits all deposits formed due to some action of glaciers e.g., BOULDER CLAY, sands and gravels occurring as fans which are carried out of the ice in glacial streams, and other deposits in the form of DRUMLINS and ESKERS.

glacial erosion the removal and wearing down of rock by glaciers and associated streams (of meltwater).

glaciation the term meaning ice-age, with all its effects, processes and products. The most recent is associated with the Pleistocene (see APPENDIX 5) but the rock record indicates older glaciations from the Precambrian and Permo-Carboniferous, and other periods in geological history.

glacier an ice mass of enormous size, usually moving. Three kinds can be identified based upon their occurrence and form, namely PIEDMONT GLACIERS, valley glaciers and ice-sheets (e.g. Greenland). An alternative classification is based on temperature and covers polar glaciers (e.g. in the Antarctic) with very low temperatures and which move slowly, primarily by deformation within the body of ice; temperate glaciers (e.g. in the Alps) where movement occurs largely by slip at the base; and subpolar glaciers (e.g. Spitzbergen) which are a mix of the two types.

globular cluster a spherical arrangement of stars, closely packed together, and containing up to many millions of stars. About one hundred clusters occur in the MILKY WAY, throughout the GALACTIC HALO.

globule a nebula made up of opaque dust and gas which denotes an early stage of star formation.

gluconeogenesis a major biochemical process occurring mainly in the liver, in which glucose is made from non-carbohydrate starting material in conditions of energy starvation. Glucose is required by red blood cells and is the primary energy source of the brain. However, the glucose reserves present in body fluids are sufficient to meet the body's needs for only about one day. Therefore, gluconeogenesis is very important during longer periods of starvation or during periods of intense muscle exercise. The raw materials used in these conditions are glycerol which is derived from the breakdown of fat; amino acids which come from proteins and lactic acid which is formed by actively contracting muscle when there is an insufficient supply of oxygen. (It is also produced by red blood cells).

glucose the most abundant naturally occurring sugar, which has the general formula $C_6H_{12}O_6$. Glucose is distributed widely in plants and animals and is an important primary energy source, although it is usually converted into polysaccharide (see SACCHARIDES) carbohydrates, which serve as long-term energy sources. The storage polymers of plants and animals are starch and GLYCOGEN respectively. Other polysaccharides of glucose include chitin and cellulose, which have a structural role and also provide strength.

glycerol a viscous, sweet-smelling alcohol, which has the chemical formula $HOCH_2CH(OH)CH_2OH$. Glycerol is widely distributed in plants and animals as it is a component of stored fats. During metabolism, stored fats break down to form the original reactants, glycerol and FATTY ACIDS, while a large amount of energy is released. Glycerol is used in industry to manufacture a range of products, including explosives, resins, toilet preparations and foodstuffs.

glycogen often called animal starch, is a polysaccharide of GLUCOSE units which occurs in animal cells (especially the muscle and liver) and acts as a store of energy released upon hydrolysis. Glycogen is also found in some fungi.

glycolysis a major metabolic process, occurring in the CYTOPLASM of virtually all living cells, where the breakdown of glucose into simple molecules generates energy in the form of ATP. Each 6-carbon glucose molecule is converted into two molecules called pyruvate, each with three carbons ($CH_2COCOOH$). This happens in a sequence of ten reactions, giving a net gain of two ATP molecules. The whole reaction sequence is regulated by several ENZYMES. Although the reactions converting glucose to pyruvate are very similar in all living organisms, the fate of pyruvate is variable. In AEROBIC organisms, pyruvate enters the MITOCHONDRIA, where it is completely oxidized to CO_2 and H_2O in a process known as Kreb's cycle (or CITRIC ACID CYCLE). This cycle, together with glycolysis, creates 38 molecules of ATP per glucose molecule. However, if there is an insufficient supply of oxygen, e.g., in an actively contracting muscle, FERMENTATION occurs and pyruvate is converted into lactic acid, liberating only 2 ATP molecules per glucose molecule. In some ANAEROBIC organisms, such as YEAST, pyruvate is converted into the alcohol ETHANOL during fermentation, again yielding only 2 ATP molecules. If a cell requires energy, or certain intermediates of the pathway are required for the synthesis of new cellular components, glycolysis proceeds, provided that glucose levels in the blood are abundant. However, when blood-glucose levels are low, e.g. during starvation, glycolysis is inhibited and instead GLUCONEOGENESIS occurs. Glycolysis and gluconeogenesis operate together such that when one process is relatively inactive, the other is highly active.

gneiss a coarse-grained METAMORPHIC rock formed during high grade regional METAMORPHISM. Gneisses usually contain QUARTZ and FELDSPAR and show banding due to the light and dark minerals separating, the dark minerals being BIOTITE mica and other silicate minerals including amphiboles or pyroxenes. Gneisses derived from SEDIMENTARY rocks are termed paragneisses, and those of IGNEOUS origin, orthogneisses.

Golgi apparatus a system of organelles within the cells of organisms, where enzymes and hormones are stored and moved. It is also involved in the formation of the cell wall where present. It is named after the Italian physician Camillo Golgi (1844–1926) who discovered its existence.

gonads the reproductive organs of animals, which produce the GAMETES and CERTAIN HORMONES. The male and female organs are known as the testes and the ovaries respectively.

graben an area of land which is bounded by two or more FAULTS with movements that result in a central block falling, relative to either side. They can form large features on the landscape and those which extend for long distances are called rifts.

graded bedding a primary feature in sedimentary rocks which shows a gradation in grain size from the base to the top of the bed. At the base is coarse-grained sand or pebbles and rising through the bed, the size diminishes through fine sand, to silt and clays on top.

gradient a measure of the steepness of a sloping line. A straight line has the equation $y = mx + c$, where x and y are the co-ordinates, c is a constant, and m is the gradient. The steepness of a point on a curve is the gradient of its tangent, which is a straight line drawn to the curve at this point.

gram (g) the basic unit of mass. There are approximately 28g per ounce, and 1000g in a kilogram.

granite a coarse-grained and commonly occurring IGNEOUS rock containing QUARTZ, alkali FELDSPAR and MICA (usually BIOTITE). Other minerals may be present including AMPHIBOLES and oxide minerals. Granites can be formed by several processes including melting of old continental crust.

graph a diagram that represents the relationship between two or more quantities, using dots, lines, bars or curves.

graphite a soft, black, variety of carbon with a hexagonal structure and a very high melting point that makes it chemically unreactive. It is a giant structure comprising a series of planes. In any one plane the bonds between the atoms are strong, but bonds between the atoms of different planes are weak and this enables the planes to slide over one another. These properties account for the fact that graphite is a good lubricant and conductor of electricity. It is used in the making of pencils and electrodes.

gravimetric analysis a method of chemical analysis in which elements are precipitated from a solution in the form of a known compound to be weighed and from which the quantity of the required element can be calculated.

gravity the attractive force that the earth exerts on any body that has mass, tending to cause the body to accelerate towards it. Other planets also exert a force of gravity, but the force is different from that exerted by the earth since it depends on the planet's mass and diameter. The true WEIGHT of any object on earth is really equal to the object's mass (m) multiplied by the ACCELERATION due to gravity (g), which is 9.8 ms^{-2}. Therefore, although weight and mass are often used synonymously, they are different for scientific purposes. For example, a man with a mass of 80kg will weigh 784 newtons (N) on earth, but on the moon he would weigh only 130N since the force of gravity on the moon is only 1/6th of that on earth. However, his mass is still 80kg and remains constant throughout the universe.

greenhouse effect the phenomenon whereby the Earth's surface is warmed by radiation from the Sun. Most of the solar radiation from the Sun is absorbed by the Earth's surface, which in turn re-emits it as INFRARED RADIATION. However, this radiation becomes trapped in the Earth's atmosphere by carbon dioxide

(CO_2), water vapour and ozone, as well as by clouds, and is re-radiated back to the Earth, causing a rise in global temperature. The concentration of CO_2 in the atmosphere is rising steadily because of mankind's activities (e.g. deforestation and the burning of FOSSIL FUELS), and it is estimated that it will cause the global temperature to rise 1.5-4.5°C in the next fifty years. Such a rise in temperature would be enough to melt a significant amount of polar and other ice, causing the sea level to rise by perhaps as much as a few metres. This could have disastrous consequences for coastal areas, in particular, major port cities like New York.

groundwater water contained in the voids within rocks. It usually excludes water moving between the surface and the WATER TABLE (*vadose* water) but may be METEORIC or JUVENILE water.

group the vertical columns of elements in the PERIODIC TABLE. Each group contains elements that have similar properties, including the same number of electrons in the outer energy level (shell). This number is represented by the group number. For example, the alkali metals in group 1 all have one electron in the outer shell, whereas the HALOGENS in group 7 all have 7 electrons in the outer shell.

guanine ($C_5H_5N_5O$) a component of the nucleic acids, DNA and RNA. Guanine is also present in many other biologically important molecules.

gum arabic (or acacia gum) a white, water-soluble powder, which in natural form is obtained from some varieties of acacia trees. It is a complex polySACCHARIDE and is used widely in pharmacy as an emulsifier (SEE EMULSION), and as an adhesive. It is also used in the food industry as an emulsifier and inhibits sugar crystallization.

gypsum an evaporite mineral, $CaSO_4.2H_2O$, which is highly insoluble and thus the first to precipitate out of sea water, to be followed by ANHYDRITE and HALITE. It commonly forms crystals and is very soft (2 on MOHS' SCALE). It is important industrially, being used in cements, fertilizer, fillers and plaster products.

H

Haber process the industrial process for the production of ammonia (NH_3) by the direct combination of nitrogen and hydrogen in the presence of an iron CATALYST. The process gives a maximum yield (40 per cent) using relatively low temperatures and high pressures. The Haber process is important in industrial chemistry since it is the most economic way to produce ammonia, from which fertilizers are made.

habit the term for the characteristic external shapes of crystals due to the number, shape, size, and orientation of the crystal faces. The extent to which perfect habit may be approached depends upon conditions during formation, speed of growth, impurities etc. Individual crystals of particular minerals may show typical growths. Habits include tabular (flat blades), fibrous (like thin fibres), prismatic (elongated) and acicular (needle-shaped).

habitat the place where a plant or animal normally lives, specified by particular features e.g. rivers, ponds, sea shore.

hadron see **elementary particle**.

haem see **myoglobin**.

haemoglobin an iron-containing red pigment, which is found within the red blood cells (called ERYTHROCYTES) of vertebrates and which is responsible for the transport of oxygen around the body. In actively metabolizing tissue such as the muscles, haemoglobin exchanges oxygen for carbon dioxide (CO_2), which is then carried in the blood back to the heart and pumped to the lungs, where the haemoglobin loses the CO_2 and regains oxygen.

haemophilia a genetic disorder affecting the blood, in which the lack of a vital BLOOD CLOTTING factor causes abnormally delayed clotting. Haemophilia usually occurs only in males, who receive the defective gene from their mothers. A haemophilic female can only arise if a haemophilic male marries a female carrying the gene (extremely rare). There is no known cure for haemophilia, and, when injured, haemophiliacs must rely on blood transfusion to replace the blood loss, which is considerably greater than that lost by a normal individual.

haemostasis see **blood clotting**.

hail hard balls or pellets of ice usually associated with CUMULONIMBUS cloud. The hail (stones) form by rain drops being taken into higher, colder regions. Upon falling the hailstone grows by the build-up of layers due to condensation of moist air on the cold hail.

half-cell as implied, half of an electrolytic cell (see ELECTROLYTE). The cell comprises an ELECTRODE dipped into an electrolyte and the potential of this system is measured against a hydrogen electrode which is given a potential of zero. A hydrogen electrode consists of hydrogen bubbling over a platinum electrode which is covered by dilute acid to produce a standard concentration of hydrogen IONS.

half-life (t) the time taken for a radioactive ISOTOPE to lose exactly half of its RADIOACTIVITY. The half-life is constant for a particular isotope, varying from a fraction of a second to millions of years, and is best determined by using a Geiger-counter (GEIGER TUBE) attached to a ratemeter which shows the count-rate. For instance, if an isotope has a half-life of one minute, then the radioactive count will fall by one half in one minute, by one quarter in two minutes, by one eighth in three minutes, and so on. Typical half-lives are: thorium 230, 80 000 years; Radium 226, 1620 years; Sodium 24, 15 hours; Radon 220, 52 seconds.

halide a compound consisting of a HALOGEN and another element. If the halogen bonds with a metal that forms a positive ion, the bond formed is ionic in nature, e.g. sodium bromide (NaBr). Halides formed by less electropositive metals and non-metals have COVALENT BONDS.

halite (or rock salt, NaCl) a common mineral in EVAPORITE deposits and often associated with GYPSUM and anhydrite. Considerable thicknesses may occur and can be mined or extracted as brine. Large masses can form 'plugs' which arch into overlying sedimentary rocks to create traps, that contain oil and gas in a rock reservoir.

halo bright ring(s) which may be seen around the Sun or Moon. It is due to REFRACTION of light by the crystals in high CIRRUS cloud.

halogen any of five elements, found in group 7 of the PERIODIC TABLE, that are the extreme form of the non-metals. They exhibit typical non-metal characteristics, existing as molecules of two atoms that are held together with COVALENT bonds, e.g. F-F (a fluorine molecule). At room temperature, fluorine and chlorine are gases, bromine is a volatile liquid, and iodine is a volatile solid. The halogens are found in nature as negative IONS in sea water and as salt deposits from dried-up seas.

halophilic bacteria bacteria which can tolerate salt and live in the surface layers of the sea. They are instrumental in various biochemical cycles including the nitrogen and carbon cycles.

halophyte a plant that can tolerate a high level of salt in the soil. Such conditions occur in salt marshes, tidal river estuaries (and on motorway verges and central reservations!) and a typical species is rice grass (*Spartina*).

hanging valley a tributary valley of a stream situated above the major river valley, possibly with a waterfall connecting the two. It is formed during glaciation and is due to the greater erosion of the main (trunk) valley by its GLACIER. The main glacier also cuts off the ends of the land between adjacent hanging valleys—creating features called "truncated spurs."

haploid this term describes a cell nucleus or an organism that possesses only half the normal number of CHROMOSOMES, i.e. a single set of unpaired chromosomes. This is characteristic of the GAMETES and is important at fertilization as it ensures the DIPLOID chromosome number is restored. For example, in a man there are 23 pairs of chromosomes per somatic (i.e. all cells but the reproductive ones) cell, which is the diploid number, but the gametes possess 23 single chromosomes, which is the haploid number.

hardness of minerals see **mineralogy**.

hard water water that does not readily form a lather with SOAP. This is due to dissolved compounds of calcium, magnesium and iron. Use of soap produces a scum which is the result of a reaction between the FATTY ACIDS of the soap and the metal ions. The scum is made up of SALTS, which when removed leave the water soft. There are two types of hardness. Temporary hardness is created by water passing over carbonate rocks (e.g. limestone or chalk), producing hydrogen carbonates of the metals, which dissolve before the water reaches the mains supply. Boiling the water decomposes the hydrogen carbonates into carbonates (producing kettle fur), and the water becomes soft. Permanent hardness in water is due to metal sulphates, which can be removed by the addition of sodium carbonate. ZEOLITES will remove both types of hardness.

Harvard classification a method for classifying the spectra of stars in which stars are called O, B, A, F, G or K type stars. A catalogue was compiled at Harvard College Observatory and is now called the Henry Draper Catalogue.

heart the hollow, muscular organ that acts as a pump to circulate blood throughout the body. The heart lies in the middle of the chest cavity between the two lungs. It is divided into four chambers, known as the right and left ATRIA, and the right and the left VENTRICLES. In normal persons there is no communication between the right side and the left side of the heart, thus the two sides act as independent pumps, which are connected in series. Starting from the left ventricle, the flow of blood is as follows:

(1) Left ventricle contracts and oxygenated blood is pushed into the AORTA under pressure.

(2) Aorta divides into numerous ARTERIES to supply blood to all parts of the body.

(3) Deoxygenated blood returning from the body is carried by small VEINS, which eventually join up to form two large veins, called the superior VENA CAVA and the inferior vena cava.

(4) These two large veins empty into the right atrium.

(5) The blood passes from the right atrium to the right ventricle via a VALVE.

(6) The right ventricle contracts, pushing blood under pressure into the PULMONARY ARTERY.

(7) The pulmonary artery branches into two, carrying blood to both the right and left lungs.

(8) Within the lungs, gas exchange occurs—carbon dioxide is expelled, and the blood is oxygenated (see HAEMOGLOBIN).

(9) The blood flows from the left atrium via a valve into the left ventricle.

The heart of other vertebrates is basically similar but invertebrates show great variation in the heart and its function.

heat ENERGY produced by molecular agitation. Heat, or thermal energy, is the kinetic and potential of the molecules in an object. The molecular movement rises with temperature, causing a rise in thermal energy.

heat capacity (C) the quantity of heat required by a substance or material that will raise its temperature by one degree KELVIN (or one Celsius). Thus, heat capacity is measured in joules per Kelvin (JK^{-1}) or joules per Celsius (JC^{-1}). The molar heat capacity of a substance is the heat required that will raise the temperature of one MOLE of the substance by one degree, and the specific heat capacity is the heat capacity per kilogram ($JK^{-1}kg^{-1}$) or gram ($JK^{-1}g^{-1}$).

heat exhaustion a physical state experienced by warm-blooded animals in which the body's normal cooling processes fail to operate as a result of increasing environmental temperature. Instead, the body's metabolic rate increases, raising the body temperature higher, which in turn raises the metabolic rate even higher, and so on. The symptoms of heat exhaustion are cramp and dizziness, and death ensues when the body temperature reaches about 42°C, which is the upper lethal temperature for the average human being.

heat of combustion the amount of heat generated when one MOLE of a substance is burned in oxygen.

heat of dissociation the amount of heat required to dissociate (see DISSOCIA-TION) one MOLE of a compound.

heat of formation the heat required or given out when one MOLE of a substance is formed from its elements (at one atmosphere and usually 298K).

heat of reaction the amount of heat absorbed or given out for each MOLE of the reactants. If the reaction is EXOTHERMIC, the convention is to specify the quantity as a negative figure (in kilojoules).

heat of solution the amount of heat absorbed or given out when one MOLE of a substance is dissolved in water.

hela cell a particular cell variety, discovered in a woman with cervical carcinoma (a form of CANCER) in 1951. These transformed cells are immortal and are used in laboratories worldwide for research purposes.

helical rising (or setting) when a star or planet rises (or sets) at the same time as the Sun.

heliometer apparatus for measuring the diameter of the SUN or the angular distance between two objects which are very close together.

heliostat an astronomical instrument, similar to the COELOSTAT, which enables study (both by photography and spectroscopy) of the Sun to be made.

helium is one of the INERT GASES (or noble gases), so called because it has a stable electronic configuration and no chemical reactivity as such. It occurs naturally in very small quantities and being non-inflammable is used as an inert atmosphere for arc welding, for airships and balloons, in gas lasers and with oxygen as the atmosphere for deep sea divers. Helium liquefies below 4K (-269°C) and is used extensively in *cryogenics*—the study of materials at very low temperatures.

Helium has no colour, taste or smell and was named after *helios*, the Greek for Sun. It is formed in stars such as the Sun as hydrogen nuclei are pressed together in the processes of nuclear fission.

helium stars SPECTRAL TYPE B stars which show dark lines due mainly to helium.

helix a curve in the form of a spiral, which encircles the surface of a cone or cylinder at a constant angle.

hemisphere a half sphere, formed when a plane is passed through the middle of a sphere, cutting it in two.

hepatitis a serious disease of the liver, of which there are two types:
    (1) Hepatitis A (infectious hepatitis) caused by the hepatitis A virus
    (2) Hepatitis B (serum hepatitis) caused by the hepatitis B virus
    Both diseases share the same symptoms of fever, nausea and JAUNDICE, but

they are transmitted by different routes. Hepatitis A is spread by the oral-faecal route and occurs in people who have poor sanitation and personal hygiene. The virus can be transmitted from person to person in contaminated food or drinking water. Most people exposed to the disease can be protected by PASSIVE immunization, which involves the administration of purified ANTIBODIES from a previously infected individual who has recovered. Hepatitis B is spread through blood products, contaminated syringes and instruments. Susceptible groups include those who require blood or blood products, e.g. haemophiliacs (see HAEMOPHILIA) although any donated blood is normally screened for hepatitis. A significant percentage of hepatitis B sufferers develop cancer, and the virus is thought to be a contributory factor.

heptagon a polygon with seven sides, the interior angles of which when added together give 900°. If the heptagon is regular, then all the sides are equal.

heredity the passing on of characteristics from parents to offspring which is accomplished through the transfer of genes (see CHROMOSOMES and GENES). Some of the basic laws of heredity were studied and worked out by an Austrian monk, Gregor Mendel (1822–1884) who carried out experiments with pea plants. He noticed that some characteristics were dominant over others, e.g. tallness is dominant to shortness in these plants. The study of genetics during this century has established that there are *dominant* and *recessive* genes for many characteristics and this is particularly important in some inherited diseases. A dominant gene is one which will always be seen in the offspring. A recessive gene will only be seen if it is present as a "double dose", i.e. one from each parent. See also DNA, GENETICS.

hertz (Hz) the unit of FREQUENCY, equivalent to one cycle per second. For example, if an OSCILLATION has a frequency of 6Hz, this means that six complete cycles occur in one second.

Hertzsprung-Russell diagram a graphical means of correlating star data in which LUMINOSITY is plotted against SPECTRAL TYPE, first used by H. N. Russell in 1913. These features relate to, and the graph is basically a plot of, total energy against surface temperature. The outcome is that most stars form a band across the graph from hot, bright, blue-white stars to cool, dim, red stars at the two extremes. This band is the MAIN SEQUENCE containing normal stars burning hydrogen. As stars evolve they move away from the main sequence, firstly as RED GIANTS. The diagram is fundamental to theories of stellar evolution.

heterocyclic compounds organic compounds forming a ring structure with the additional elements, e.g. oxygen, hydrogen, nitrogen and sulphur.

heterozygote an organism having two different ALLELES of the GENE in question in all somatic (non-reproductive) cells. For instance, if gene B has two allelic forms, B and b, then the heterozygote will contain both alleles, i.e. Bb, at the appropriate location on a pair of HOMOLOGOUS CHROMOSOMES. Heterozygotes can thus produce two kinds of GAMETES, B and b. One allele of a heterozygote is usually dominant, and the other is usually recessive. The dominant allele is the one that is expressed in the characteristics of the organism, because it masks the expression of the recessive allele. Dominant alleles are usually denoted by capital letters, while recessive alleles are denoted by lower case letters (see *also* HEREDITY).

hexagon a polygon with six sides. If the hexagon is regular, then each of the interior angles will be 120°.

hiatus a break in a succession of sedimentary rocks due to erosion or non-deposition.

Hill reaction the light-dependent stage in the process of PHOTOSYNTHESIS, in which illuminated CHLOROPLASTS initiate the splitting of water. This produces hydrogen atoms (two per water molecule), which are used to reduce carbon dioxide with the formation of carbohydrate in the dark stage of photosynthesis. The light stage also produces ATP, which provides the energy required for carbohydrate synthesis.

histogram a graph that represents the relationship between two variables using parallel bars, but it differs from a bar chart in that the frequency is not represented by the bar height, but by the bar area.

histamine an AMINE released in the body during allergic reactions, and in injured tissues. Release causes dilation of blood vessels, causing a fall in blood pressure.

histology the study of the tissues, tissue structure and organs of living organisms in the main through microscopic techniques.

histolysis breakdown of a cell or tissue.

HIV (*abbreviation for* human immune deficiency virus) the retrovirus thought to be the cause of AIDS.

hoar frost ice crystals formed on surfaces, e.g. vegetation, cooled by heat loss through radiation. The ice comes from frozen dew and the SUBLIMATION of water vapour to ice.

hole see **electric current**.

holography a method of recording and reproducing three-dimensional images using light from a LASER, but without the need for cameras or lenses. The holographic images are generated by two beams of laser light producing interference patterns. A single beam of laser, or coherent, light is split into two. One beam is reflected onto the object and then onto the photographic film or plate. The second reference beam passes straight onto the film. The interference pattern on the film produces a **hologram**. The developed film, when illuminated by coherent light, reproduces the image because the interference patterns break up the light, which reconstructs the original object. Because a screen is not required, the light forms a three-dimensional image in air.

homeostasis the various physiological control mechanisms that operate within an organism to maintain the internal environment at a constant state. For example, homeostasis operates to keep the body temperature of humans within a small, crucial temperature range, independent of the temperature of the external environment, as our metabolic processes would not function in any other temperature range.

homologous the term given to organs or structures that have evolved from a common ancestor, regardless of their present-day function. For example, the pentadactyl limb is the ancestral form of the quadruped forelimb, and from it evolved the human arm, the fin of cetaceans, and the wings of birds. These structures are therefore said to be homologous. Similarities in homologous structures are best seen in early embryonic development and imply relationships between organisms living today.

homologous series chemical compounds that are related by having the same functional group(s) but formulae that differ by a specific group of atoms. For instance, the ALKENES form an homologous series in which each successive member has an additional CH_2 group, i.e.:

Alkene Series	Molecular Formula
Ethene	C_2H_4
Propene	C_3H_6
Butene	C_4H_8

homozygote an organism that has two identical ALLELES of the GENE in question in all SOMATIC CELLS. For instance, if gene B has two allelic forms, B and b, the homozygote will contain only one allelic type, i.e. either BB or bb, at the appropriate location on a pair of HOMOLOGOUS CHROMOSOMES. Homozygotes can thus only produce one kind of gamete, B or b, and as such are capable of what is called pure breeding. For example, the gene for albinism is RECESSIVE, and any individual that possesses this characteristic will be homozygous for that gene. If two such individuals breed, the resultant offspring will all be albinos.

Hooke's law the physical relationship between the size of the applied force on an elastic material and the resulting extension. The extension must be within the YIELD zone of the material, as any force that goes beyond this will cause permanent deformation. Hooke's law can be represented by the equation $T = kx$, where T is the magnitude of force, k is the spring constant, and x is the displacement of the material.

horizon in geology, a general term used in stratigraphy referring to a plane within a series of rock layers. It has little or no thickness as such and is used to pinpoint changes in rock type or may refer to a very thin bed within a unit.

In astronomy, the great circle, with the NADIR and ZENITH forming the poles, which marks out a horizontal plane where it meets the CELESTIAL SPHERE.

horizontal the term in mathematics that describes a line that is at right-angles with the vertical and parallel to the horizon.

hormone an organic substance, secreted by living cells of plants and animals, that acts as a chemical messenger within the organism. Hormones act at specific sites, known as "target organs," regulating their activity and producing an appropriate response. In animals, hormones are secreted from various ductless glands, which include the pancreas, thyroid and adrenal glands. This hormone-signalling system is collectively known as the ENDOCRINE SYSTEM. These glands secrete hormones directly into the bloodstream, usually in small amounts, where they circulate until they are picked up by appropriate receptors present on the cell membranes of the target organs. These receptors recognize the particular hormone and bind to it, producing the response. Hormones also play an important part in the role of plant and seed growth and are found in root tips, buds, and other areas of rapid development. For example, gibberellins are a class of plant hormones involved in processes such as initiating responses to light and temperature, the formation of fruit and flowers, and the promotion of shoot elongation. Hormone action is constantly regulated by elaborate FEEDBACK MECHANISMS, both within and between cells and organs, to regulate the secretion and breakdown of hormones. One example of a hormone in the human body is insulin, which is produced by the pancreas and controls the level of sugar in the blood.

Horsehead nebula a well-known DARK NEBULA in the Orion CONSTELLATION.

horst an area of land which is bounded by two or more FAULTS with opposite senses of movement, resulting in a central block which is raised, relative to either side.

hot-wire anemometer a device for measuring the speed of the wind. It comprises a current-carrying wire which is cooled by the wind and calculations can be made to relate the wind-speed to the RESISTANCE of the wire.

hovercraft see **friction**.

Hubble constant a measure of the rate of expansion of the universe and its variation with distance. It is, in effect the time period since all matter was together in one mass. It is calculated from the RED SHIFTS of distant GALAXIES and produces figures of between 5 and 10 thousand million years.

humidity the amount of water vapour in the earth's atmosphere. The actual mass of water vapour per unit volume of air is known as the absolute humidity and is usually given in kilograms per cubic metre (kgm^{-3}). However, it is useful to use relative humidity, which is the ratio, as a percentage, of the mass of water vapour per unit volume of air to the mass of water vapour per unit volume of saturated air at the same temperature.

humus material that makes up the organic part of soil being formed from decayed plant and animal remains. It has a characteristic dark colour and its composition varies according to the amount and type of material present. It holds water (and it forms a physical state known as a COLLOID), which can then be used by growing plants and helps to prevent the loss of minerals from the soil by leaching. Hence it is very important in determining soil fertility. Humus may be more ACIDIC (*mor*), as in the soil of coniferous forests, or ALKALINE (*mull*), as found in the soil of deciduous woodlands and grasslands. Humus contains numerous MICRO-ORGANISMS and INVERTEBRATE animals, and its presence is of obvious economic importance in the cultivation of food crops.

Huntington's chorea see **lethal gene**.

hurricane on the BEAUFORT SCALE, a wind with an average speed of 64 knots or more (74 mph). Also, an intense tropical cyclonic storm in which the winds circulate at high speeds. Hurricanes occur in the Pacific and North Atlantic and in the western Pacific are known as a TYPHOON.

hydrides compounds formed by the reaction of elements with hydrogen. A number of compound types are formed, distinguished by bonding and molecular structure. Hydrides of alkali or alkali earth metals are salt-like and crystalline; non-metals often form liquids or gases many of which dissolve in water forming an acid or alkali; and TRANSITION ELEMENTS form hybrids.

hydrocarbon an organic compound that contains carbon and hydrogen only. There are many different hydrocarbon compounds, the most common being the ALKANES, ALKENES, and ALKYNES. There is a fundamental division by structure, forming two groups. Aliphatic hydrocarbons are made up of a chain of carbon atoms whereas aromatic hydrocarbons have a ring structure.

hydrochloric acid an aqueous solution of hydrogen chloride gas, producing a colourless, fuming, corrosive liquid. It will react with metals to form chlorides, liberating hydrogen. It is made by the ELECTROLYSIS of brine producing hydrogen and chlorine, which are combined, or by the reaction of SULPHURIC ACID with sodium chloride. It has many uses in industry, in the manufacture of other chemicals and in food processing.

hydrogen the lightest element, which forms molecules containing two atoms—H_2. It occurs in the free state and is widely distributed as a component of water, minerals and organic matter. It is manufactured by the electrolysis of water and is also produced in the catalytic (see CATALYST) treatment of petroleum. It is explosive over a wide range of mixtures with oxygen and combines with most elements to form HYDRIDES. It has numerous uses including HYDROGENATION reactions where hydrogen is added to a substance (used extensively in the petrochemicals industry), organic and inorganic synthesis, e.g. production of methanol, ammonia and hydrochloric acid, also metallurgy, and filling balloons. It has three isotopes, protium, DEUTERIUM and tritium containing 0, 1 and 2 NEUTRONS respectively.

hydrogenation an important industrial reaction where gaseous HYDROGEN is the vehicle for adding hydrogen to a substance. The reaction usually proceeds in the presence of a CATALYST and often at elevated pressure. This reaction is utilised in the petroleum refining and petrochemicals industry, the hydrogenation of coal to produce HYDROCARBONS and the hydrogenation of fats and oils.

hydrogen I and II the hydrogen identified in space and which occurs in two forms. Hydrogen I (otherwise known as neutral hydrogen) is seen in the spiral arms of the GALAXY and emits radio radiations. Hydrogen II (or ionized hydrogen) is found in gaseous nebulae, emitting visible and radio radiations.

hydrogen bomb see **nuclear fusion.**

hydrogen peroxide (H_2O_2) a strong oxidizing and bleaching agent usually in the form of a SOLUTION in water. On decomposing it produces water and oxygen, hence it is used as a bleach. It is used industrially and recently it has been used as the oxidizing agent in rocket fuel.

hydrology the study of water and its cycle, dealing with bodies of water and how they change, and all forms of water (rain, snow and surface water) including its use, distribution and properties. It involves aspects of oceanography, meteorology and geology.

hydrolysis the term used to describe a chemical reaction where the action of water causes the decomposition of another compound and the water itself is decomposed. In salt hydrolysis, the salt dissolves in water, producing a solution that may be neutral, acidic or basic, depending on the relative strengths of the ACID and BASE of the salt. For example, a solution of potassium chloride (KCl) would be neutral, since potassium forms a strong base and chlorine forms a strong acid. In comparison, ammonium chloride (NH_4Cl) gives an acidic solution, since ammonium forms a weak base but chlorine forms a strong acid.

hydrometeor a general term encompassing all forms of water vapour in the atmosphere which have condensed or sublimed e.g. RAIN, SNOW, FOG, CLOUD and DEW.

hydrometer an instrument consisting of a bulb with a long stem which, when floated in a liquid with the stem upright, enables the relative density of the liquid to be measured.

hydrosphere the water that exists on or near to the earth's surface. The main components are water (H_2O), sodium chloride (NaCl) and magnesium chloride ($MgCl_2$). By mass, the major elements are oxygen (almost 86 per cent), hydrogen (10.7 per cent), chlorine (2 per cent) and sodium (1 per cent). Magnesium is the only other element present in significant quantities.

hydrothermal metamorphism the reaction of very hot waters, produced by hydrothermal activity (associated with igneous activity), with the rocks through which they pass resulting in the alteration of minerals.

hydroxide a compound derived from water (H_2O) through the replacement of one of the hydrogen atoms by another atom or group, e.g. NaOH, sodium hydroxide. ALKALIS are the hydroxides of metals.

hydroxyl the OH group comprising an oxygen and a hydrogen atom bonded together. In alcohols the OH group occurs in a COVALENTLY bonded form.

hyetograph apparatus for the collection and measurement of rainfall.

hygrometer an instrument for measuring the relative HUMIDITY of the air.

hygroscopic the term applied to a substance which absorbs moisture.

hypertonic when one liquid has a higher osmotic pressure (see OSMOSIS) than another, or a standard, with which it is being compared.

hypotenuse in a right-angled triangle, this is the name given to the longest side, which always faces the right angle.

I

ice age the spreading of ice as glaciers over areas that have, before then, been ice-free. Most is known about the last ice age during the Pleistocene geological epoch 2 million years ago. However, geological studies show evidence of earlier ice ages although less is known about these.

iceberg a large lump of ice which has broken off from a GLACIER or ice sheet and is floating in the sea. Icebergs can be carried long distances before they finally melt.

identical twins two identical individuals which develop from a single fertilized egg that divides early on into two equal parts. Identical twins have the same genetic make up and are always of the same sex.

igneous rock one of the three main rock types that is formed by the forcing up of hot, molten rock (magma), from great depths, or lava flows on the surface associated with volcanoes.

immiscible the term that describes liquids that cannot be mixed together such as water and oil. Two liquids which can be mixed together are described as being *miscible*.

immune system the natural defence system, present in the body of a VERTEBRATE animal which helps to protect it against diseases caused by micro-organisms and parasites. There are two types of immunity provided by the immune system:

1) *natural* (known as innate)—this is present from birth and is wide ranging, operating against almost any foreign substance that threatens the body.

2) *acquired immunity*—this type of immunity occurs as a result of an attack by a particular foreign substance. The immune system builds up its defences against that particular substance and uses them again in any future attack.

immunoglobins (Ig) these are groups of PROTEINS, known as ANTIBODIES, which are produced by special cells in the blood called B CELLS. Antibodies can join or bind on to particular foreign substances (called ANTIGENS) and make them ineffective and not able to cause disease. If particular antigens are present, the B cells divide and the multiplied cells produce quantities of immunoglobulins.

impermeable a substance which does not allow the passage of gases or liquids.

incandescence the condition in which a substance is at a high enough temperature to give out light, e.g. the filament of an electric light bulb.

independent variable in a mathematical expression, this is a variable that may have any value, which does not depend upon the value of the other quantities present. For example, in the equation $y = 3x + 8$, the value of y depends upon that of x so it is the DEPENDENT VARIABLE, but x is the independent variable.

indicator a chemical substance, usually a large, organic molecule, that is used to detect the presence of other chemicals in a solution. Indicators are usually weak acids which change colour (due to becoming ionized) depending upon the pH of the solution under investigation.

inequality the mathematical statement which shows which of two quantities is the larger or smaller. Therefore, if x is larger than y it is written as $x > y$, but $x < y$ indicates that x is smaller than y.

inert gas any one of a number of elements which make up group 8 of the PERIODIC TABLE. The inert gases are helium, neon, argon, krypton, xenon and radon. Helium is found in collections of natural gas and has a lower BOILING POINT than any other substance.

inertia the property or characteristic of an object that causes it to resist any change in its state of movement, unless it is being acted upon by some outside force. If no force is present, the object will remain still or continue moving at a particular speed in a straight line. The first of NEWTON'S LAWS OF MOTION says that the MASS of an object is directly related to the amount of its resistance to changing its direction of motion, or its inertia.

infinity (∞) the word used to describe a number or quantity with a value which is too enormous to be measured. It is usually used to describe space which is believed to be boundless and limitless and so is called infinite. Negative infinity, written $-\infty$ describes a value which is too minute to be measured and so insignificant in a calculation. Such a value is also called *infinitesimal.*

infrared astronomy the study of INFRARED RADIATION which is sent out from celestial bodies such as stars. This radiation tends to be mopped up or absorbed by water vapour in the Earth's atmosphere. Sometimes, recording apparatus and observatories are placed on the tops of mountains in order to try and tackle this problem.

infrared radiation a type of electromagnetic radiation with a particular wavelength range (from 0.75μm to 1mm) lying between that of visible light and microwaves. This radiation can travel through fog and can be detected using special recording equipment and photographic plates.

inhibitor in chemistry and biology, a substance that stops or slows down a chemical reaction.

initiator a substance that starts a chemical reaction.

inorganic chemistry the branch of chemistry that is concerned with ELEMENTS and compounds excluding CARBON.

insoluble in chemistry, a substance that will not dissolve in a SOLVENT or will only dissolve very slightly.

insulator a substance that is good at keeping in heat or one which does not allow an electric current to flow because it is a poor CONDUCTOR. Examples include plastics, rubber, polystyrene and many non-metals.

insulin an important hormone produced by an organ called the pancreas in a VERTEBRATE animal which controls the level of glucose (sugar) in the blood and cells of the body. People with diabetes mellitus have too much sugar in the blood due to a lack of insulin production by the cells of the pancreas. They need to take extra insulin in the form of an injection or tablets.

integer any number that is a whole number and not a fraction which may be negative or positive and includes zero, e.g., -4, -3, -2, -2, 0, 1, 2, 3, 4.

integration a branch of the mathematical study, called CALCULUS, using various types of formulae.

interference the meeting or interaction of two or more waves (water, electromagnetic or sound), which are passing through a particular area at the same time. Constructive interference occurs between waves that are in phase with one another, that is, their troughs and crests coincide. This results in a larger wave of greater size or AMPLITUDE being produced. Destructive interference occurs when waves are out of phase, that is, both crests and troughs overlap, resulting in a smaller wave with a lesser amplitude. If two waves are exactly out of phase then they cancel each other out. Interference in radio waves is a cause of crackling and buzzing, or poor reception.

intrusion a mass of (usually) IGNEOUS rock that is pushed into existing rocks. There are several different kinds of intrusion and some are very large and massive while others are much smaller.

inversion or **temperature inversion** the situation in which temperature rises with height through the atmosphere, instead of falling as is normally the case.

in vitro a word used to describe experimental biological or biochemical laboratory activities which are carried out 'in glass'. In other words, it describes experiments or activities carried out in dishes or flasks; as opposed to processes occurring within an organism.

in vivo a word used to describe biological and biochemical processes which occur inside living cells or organisms.

ion exchange the exchange of IONS, which have the same charge, between a

SOLUTION and an INSOLUBLE solid. This is the basis of separation techniques such as ion exchange chromatography.

ionosphere the upper part of the ATMOSPHERE, above 80km or 50 miles, which contains many IONS and unattached or free ELECTRONS.

iron (Fe) a metallic ELEMENT in group 8 of the PERIODIC TABLE. It occurs naturally in rocks as magnetite, haematite, limonite, siderite and pyrite and is extracted or separated mainly by the blast furnace process. It is the most useful and widely used of all metals.

irrational number a number that cannot be made into a fraction and can only be approximately made into a decimal. An example is the $\sqrt{2}$ which has an approximate decimal value of 1.414.

isobar a line on a weather map joining points of equal atmospheric pressure. If isobars are close together, the weather is usually stormy and changeable.

isohel a line on a map joining places that have the same amount of sunshine.

isohyet a line on a map joining places that have the same amount of rainfall.

isomer a chemical compound that has the same molecules as another chemical compound but a different arrangement of atoms. They are studied in a branch of chemistry called STEREOCHEMISTRY.

isosceles a word used to describe a triangle that has two equal sides and angles. Also, it describes a type of TRAPEZIUM where the two sides which are not parallel are equal.

isotherm a line joining places with the same temperature.

isotonic the term used to describe solutions with the same osmotic pressure (see OSMOSIS) so there would be no movement of particles or ions between them.

J

jasper an impure variety of minutely crystalline quartz (SiO_2). It is usually red, brown or yellow, and some varieties are banded, e.g. Egyptian or ribbon jasper.

jaundice a condition characterized by the unusual presence of bile pigment circulating in the blood. Jaundice is caused by the bile produced in the liver, which should go into the intestines, passing instead into the circulation because of some obstruction. The symptoms of jaundice include a yellowing of the skin and the whites of the eyes.

jet stream a high-speed westerly wind in the region of the tropopause. The principal jets are the polar front and subtropical jets, which occur at heights between 10 and 15 km. The winter jet stream occurs between 50 and 80 km in the upper STRATOSPHERE. The wind speeds can reach 300 km/h. The air streams move north and south of their general trend, in surges, which are probably the cause of depressions and anticyclones. There are a number of separate jet streams, but the most constant is that of the subtropics. Jet stream speed and location are of importance to high-flying aircraft.

joint 1 in geology, fractures in rock that may occur as parallel sets or, more commonly, in an irregular and less systematic manner. Where a set of joints can be identified, it can usually be related to tectonic (that is to do with deformation of the Earth) stresses and the geometry of the rock body. There are several types of joint: unloading joints, which are caused by the release of stress on rocks at depth as overlying rocks are removed by erosion; cooling joints, which occur in igneous bodies; and joints that are related to regional deformation (on a wide scale).

2 In biology, the point at which bones and the surrounding tissues meet. They may be fixed joints (as in the cranium), slightly movable (e.g., joints in the spine) or freely movable (e.g., knee, arm). The movable joint types also differ – ball and socket (hip) or hinge (elbow).

joule the unit of all ENERGY measurements. It is the mechanical equivalent of heat, and one joule (J) is equal to a force of one NEWTON moving one metre, i.e. $1J = 1Nm$. It is named after James Prescott Joule (1818–1889), a British physicist who investigated the relationship between mechanical, electrical and heat energy, and, from such investigations, proposed the first law of THERMODYNAMICS, the conservation of energy.

Julian calendar the calendar introduced by Julius Caesar. Based on a year of 365.25 days, it remains the calendar in use today although it is modified.

Julian date a system for the consecutive numbering of days, irrespective of month and year, which is used especially in astronomy. The starting point was midday on 1 January, 4713 BC, and the system was introduced by J. Julius Scaliger in 1582. (There is no connection with the Julian calendar).

Jupiter the giant planet of our SOLAR SYSTEM, more than one thousand times larger than Earth. It is one of the 'gas giants', being mainly composed of hydrogen. Its atmosphere is made up of hydrogen with approximately 15 per cent helium and traces of water, ammonia and methane. This forms a liquid 'shell' surrounding a zone of metallic hydrogen (that is, the hydrogen is compressed so much that it behaves like metal), which itself surrounds a core partly made of rock and ice. This core has a mass ten times greater than that of the Earth. Violent storms and winds rage around Jupiter, whipping up bands of frozen chemicals such as ammonia. One such storm is the *Great Red Spot*, visible on the surface as an enormous cyclone that has probably lasted for hundreds of years. Jupiter spins very rapidly so that one of its days lasts for only nine hours and fifty minutes. This rapid spin drags the whirling gases into bands that appear dark and light. A year on Jupiter lasts for nearly 12 Earth years because the planet is farther from the Sun and has a greater orbit. Jupiter is the fifth planet from the Sun, and because of its rapid rate of spin, it bulges outwards at its equator. Hence the diameter at the equator is 142,800 km, compared to 134,000 km at the poles. The outermost layers of Jupiter are very cold, in the region of -150°C, but the very centre of the planet is extremely hot, probably exceeding the temperature of the Sun. Jupiter's great mass means that it exerts a strong gravitational pull and is able to hold down the molecules of gas which swirl around its bulk. A person on Jupiter would be twice as heavy compared to his or her weight on Earth.

Jupiter has its own satellites, or moons, orbiting around it, and some of these are as large as Earth's moon. The *Voyager* spacecraft passed close to Jupiter in 1979 and sent back to Earth fascinating information about the planet and its moons. This revealed that two of the moons, called *Ganymede* and *Callisto*, have craters pitting their surface, like Earth's moon. Another moon, *Europa*, was shown to be a ball of yellow ice. The closest moon to Jupiter, *Io*, has several erupting volcanoes and a surface of yellow sulphur. Enormous electrical energy exists between Io and Jupiter, estimated to be equivalent, in any one second, to all the electrical power generated in the United States.

juvenile water water that originates from a MAGMA and has never been in the atmosphere. Surprisingly, great quantities of water can originate in this way.

K

kame a structure produced by glacial deposition (see GLACIAL DEPOSITS). It occurs as a mound of sands and gravels, with bedding (that is, it is arranged in layers) and often slumping at the sides. It was formed by the melting of stagnant ice which as a result dropped its load of sediment.

kaolin see **china clay** and **kaolinite**.

kaolinite aluminium silicate with water in its structure which occurs as an alteration product in IGNEOUS rocks, GNEISSES and a weathering and hydrothermal reaction product in sedimentary rocks. It is used in ceramics and as a coating and filler in the paper industry.

kaolinization the high temperature hydrothermal alteration and breakdown of FELDSPARS to form KAOLINITE. The process can proceed in granites to the point where the rock literally falls apart and the only recognisable, original mineral is quartz (see CHINA CLAY).

karst a karst landscape is one created in a limestone area and which occurs due to the limestone itself. The distinctive landforms are a result of solutions moving through joints and fissures in the rock dissolving the rock, enlarging cracks and joints and creating underground waterways and caves. Features typical of a karst include networks of furrows and sharp crests, funnel-shaped hollows and eventually conical hills and steep-sided depressions of impressive scale.

karyotype the number, shapes and sizes of the chromosomes within the cells of an organism. Every organism has a karyotype that is characteristic of its own species, but different species have very different karyotypes. For example, all normal human females have 22 pairs of DIPLOID chromosomes with a similar shape and size, but all female horses have 32 pairs of diploid chromosomes which have their own unique shape and size.

katabatic wind the sinking and downward movement of cold, dense air beneath warmer, lighter air. The air is cooled by radiation, usually at night. It occurs over ice-covered surfaces and in the fjords of Norway, and in many cases can be gale force.

kelp a large, brown, algal seaweed found anchored to the sea bed below low tide level. Kelp is a source of iodine and potash.

Kelvin scale the unit of temperature (K) based on the temperature scale devised by the British physicist Lord William Kelvin (1824–1907). The Kelvin scale has positive values only, with a lowest possible value of 0K, which is equal to -273.15°C or -459.67°F.

Kepler's laws of planetary motion a set of laws named after Johannes Kepler and published early in the 17th century. Kepler's laws state that the planets move about the Sun in ellipses with the Sun at one of the foci of the ellipse (an ellipse has two foci, or 'centres'); a line joining the Sun to each planet covers equal areas in equal times; the square of the planet's year and the cube of its average distance of the Sun is the same proportion for all planets.

keratin a fibrous, sulphur-rich protein consisting of coiled POLYPEPTIDE chains which occur in hair, hooves, horn and feathers.

kerosene (or **kerosine**) a thin oil that is one of many products obtained during the FRACTIONAL DISTILLATION of PETROLEUM. Kerosene is used as fuel for jet engine aircraft.

ketone an organic compound that contains a C=O (carbonyl) group within the compound rather than at either end of the compound. There are many forms of ketones, and their physical and chemical properties differ due to the presence of alkyl groups ($-CH_3$) or aryl groups ($-C_6H_5$) within the ketone molecule. Ketones can be detected within the bodies of humans when fat stores are metabolized to provide energy if food intake is insufficient. If the ketones accumulate within the blood, the undernourished individual will then experience symptoms such as headaches and nausea. The presence of ketones in urine is called ketonuria.

ketone body one of the compounds produced by the liver due to metabolism of fat deposits in the body. In abnormal conditions when the supply of carbohydrates is reduced (due to starvation or diabetes) the level of ketone bodies in the blood rises and may be present in the urine giving off a 'pear-drops' odour due to the presence of acetone.

kettle hole a hole or depression formed in glacial drift (deposits) due to outwash material from a GLACIER covering isolated masses of ice. When the covered ice eventually melts, the sediments slump down into the space creating a surface depression.

kidney one of a pair of organs responsible for removing waste products from the blood. There is a pair of kidneys at the back of the abdomen and each contains numerous tubules called nephrons which expand at one end forming the Bowman's capsule. Blood enters the capsule and water and wastes pass along the nephron where useful compounds (e.g., water) are reabsorbed. The waste liquid left (urine) empties into the bladder.

kilobyte (KB) see **byte**.

kilocalorie a unit of heat used to express the energy value of food. One kilocalorie is the heat needed to raise the temperature of one kilogramme of water by 1°C. It is estimated that the average person needs 3000 kilocalories per day, but this requirement will vary with the age, height, weight, sex and activity of the individual.

kilogram a unit of mass (kg) that is equal to the international prototype made of platinum and iridium stored in the French town of Sèvres.

kinesis the response of an organism to a particular stimulus in which the response is proportional to the intensity of the stimulation.

kinetic energy the energy possessed by a moving body by virtue of its mass (m) and velocity (v). The kinetic energy (Ek) of any moving body can be determined using the following equation:

$$Ek = \tfrac{1}{2}mv^2 \text{ (the energy is in joules if } m \text{ is kg and } v \text{ ms}^{-1})$$

As the kinetic theory of matter states that all matter consists of moving particles, it holds that all particles must possess some amount of kinetic energy, which will increase or decrease with the surrounding temperature (see EVAPORATION).

knot a unit of nautical speed equal to one NAUTICAL MILE (1.15 statute miles or 1.85 kilometres) per hour. The term knot originates from the period when sailors calculated their speed by using a rope with equally spaced tied knots, attached to a heavy log trailing behind the ship. The regular space between knots was measured at 47 feet, 3 inches, which is 14.4 metres.

Köppen classification a system of climatic classification developed between 1910 and 1936 based upon annual and monthly means (averages) of temperature and precipitation, and the major types of vegetation. The system has three orders or levels beginning with the overall climate e.g., warm, temperate and rainy (class C), which can then be divided further as for example - winter, dry produces class Cw. The third level qualifies temperature, for example a hot summer would produce a classification Cwa.

The major climates are: tropical rainy (A), dry (B), warm temperate rainy (C), cold snowy (D) and polar (E).

Kreb's cycle see **citric acid cycle**.

L

labelled compound a compound used in radioactive tracing, where an atom of the compound is replaced by a radioactive ISOTOPE, which can be followed through a biological or physical system by means of the RADIATION it emits.

labile a term used in chemistry which means not stable, likely to change.

laboratory any room or building that is especially built or equipped for undertaking scientific experiments, research or chemicals manufacture. The study of physics, chemistry, biology, medicine, geology and other subjects usually involves some work in a laboratory and each is equipped with special instruments.

A chemical laboratory typically contains balances for weighing samples, bottles of chemicals including dangerous acids, a vast array of glassware (test tubes, flasks etc.) and bunsen burners. The *bunsen burner* is a gas burner consisting of a small vertical tube with an adjustable air inlet at the base to control the flame. The flame produced by burning the hydrocarbon gas/air mix has an inner cone where carbon monoxide is formed and an outer fringe where it is burnt. When the gas is burnt completely the flame temperature is very high, around 1450°C. The burner was invented by Robert Wilhelm Bunsen, a German scientist.

Lagrangian points in particular, with respect to the Earth and Moon, the locations where the gravitational forces are equal and so objects positioned at such points remain fixed. There are five such points between the Earth and Moon.

land and sea breezes air circulation along coasts during summer, developed when the overall pressure gradient is at its lowest. During the day, the Sun warms the land more than the adjacent sea and so the air above the land becomes warmer and thus rises. This produces a convective motion with the cooler air from the sea moving onto the land. At night the situation is reversed because the sea is warmer than the land.

Langmuir's theory the basic idea which was basically correct, that electrons in an atom are arranged in shells which correspond to the periods of the PERIODIC TABLE, and that the most stable configuration (structure) is a complete shell.

lanthanides (*otherwise known as the* **rare earth elements**) these elements, from lanthanum (La) to lutetium (Lu), have much in common, chemically, with the scandium group (group 3B of the PERIODIC TABLE). The properties of these metals are very similar, and the lanthanides and yttrium (symbol Y) all occur together and can separated by CHROMATOGRAPHY. The elements are reactive,

with the heavier ones resembling calcium, while scandium is similar to ALU-
MINIUM.

laser is an acronym (abbreviation) for *L*ight *A*mplification by *S*timulated *E*mission
of *R*adiation and is a device that produces an intense beam of light of one wavelength (called *monochromatic*) in which the waves are all in step with each other
(*coherent* light). In its simplest form a ruby crystal shaped like a cylinder is subjected to flashes of white light from an external source. The chromium atoms in
the ruby become excited through absorbing photons of LIGHT and when struck
by more photons, light energy is released. One end of the cylinder is mirrored
to reflect light back into the crystal and the other end is partially reflecting, allowing the escape of the coherent light. The ruby laser produces pulses of laser
light and is called a *pulse laser*.

Lasers have also been constructed using INERT GASES (helium and neon
mixed; argon alone) and carbon dioxide. These are called *gas lasers*, and produce a continuous beam of laser light. Lasers are being used for an increasing
number of tasks including printing, communications, compact disc players,
cutting metals, HOLOGRAPHY and for an ever-widening range of surgical techniques in medicine. Lasers are also used in shops at the checkout, to read
bar codes.

Lassaigne's test a chemical test for the presence of nitrogen and also SULPHUR
or HALOGENS. The sample is heated with sodium in a test tube, quenched and
ground and on reaction with certain reagents a characteristic product/colour is
produced.

latent heat the measurement of heat ENERGY involved when a substance changes
state. While the change of state is occurring, the gas, liquid or solid will remain
at constant temperature, independent of the quantity of heat applied to the
substance (an increase in heat will just speed up the process). The specific latent
heat of fusion is the heat needed to change one kilogram of a solid into its liquid
state at the MELTING POINT for that solid. For example, the specific latent heat of
fusion for pure, frozen water (ice) at 273K (0°C) is 334 kJkg^{-1}. The specific latent heat of vaporization is the heat needed to change one kilogram of the pure
liquid to vapour at its boiling point. In the case of pure water again, at its boiling
point of 373K (100°C), 2260kJkg^{-1} is the specific latent heat of vaporization
needed to change water into steam.

lateral line the sensory system of fish and aquatic amphibians which consists of
sensory cells called neuromast organs in a line along the body. Movements in
the water affect the neuromasts thus creating nerve impulses.

lateral moraine rock debris created by a GLACIER which accumulates at the margin of a valley glacier and is due to transport and reworking of rocks from the valley sides which eventually accumulates as MORAINE.

latex originally used to describe the fluid obtained from rubber trees which is essentially a SUSPENSION of rubber in water. It occurs as a milky fluid containing mineral salts, sugars, proteins, oils and alkaloids and coagulates on exposure to air. The term is also applied to synthetic polymers and rubbers which are produced as latexes and which may be used for the manufacture of goods.

latitude the angular distance of a particular point on the Earth's surface measured relative to the Earth's equator. Latitude is measured in degrees corresponding to the angle of incident light from a specific star, e.g. the Sun, above the horizon at a given time and is described as being north or south of the equator. On a world globe, lines of latitude are represented by parallel, horizontal lines. The equator is 0° while the poles are 90° north or south.

lattice the particular arrangement of atoms in a CRYSTAL structure. The smallest, complete lattice, which is repeated throughout the crystal, is called the unit cell.

lattice energy the energy needed to break down one MOLE of a substance from its crystal LATTICE into the gaseous state where the constituent ions are very far apart.

lava molten rock erupted by a VOLCANO, whether on the ground (subaerial) or on the sea floor (submarine). The chemical composition varies and they show a number of textures but all are characterized by some glass-like material and/or fine-grained minerals, because they have cooled so rapidly that large crystals could not form. The way it is erupted and moves and therefore its subsequent shape and form depends greatly on the VISCOSITY, and generally a less viscous lava will flow faster. Two types of basaltic (see BASALT) lava forms are seen: 'aa' which looks jagged and 'pahoehoe' which exhibits a smoother ropy appearance.

law of conservation see **energy, thermodynamics**.

Le Chatelier's principle a statement relevant to chemical reactions, which predicts that if the conditions of a system in EQUILIBRIUM are changed, the system will attempt to reduce the enforced change by shifting equilibrium.

lee wave a stationary (STANDING) wave in air in the lee of a mountain barrier, created by air passing over the mountain and then returning to its original level. Sometimes such waves may have large AMPLITUDES and clouds often form along the wave crests.

legionnaires' disease an infectious disease caused by the BACTERIUM *Legionella pneumophila*, which inhabits surface soil and water. It has also been traced in water used in air-conditioning cooling towers. The main source of infection is inhalation of air or water carrying the bacteria, and so far there is no evidence that it is transmitted from an infected to a non-infected individual. Legionnaires' disease is really a form of pneumonia, and thus its symptoms include shortness of breath, coughing, shivering and a rise in body temperature. Healthy individuals should fully recover from infection if treated with the antibiotic called erythromycin.

lens is a device that makes a beam of rays passing through it either converge (meet at a point) or diverge. Optical lenses are made of a uniform transparent medium such as glass or a plastic and they refract the light (see REFRACTION). They are either convex (thickest in the middle) or concave (thickest at the edges) in shape and lenses can be made with combinations of these profiles or one side may be flat (plane). Light rays that pass through a convex lens are bent towards the *principal axis* (or optical axis—the line joining the centre of curvature of the two lens surfaces), and away from the axis with a concave lens. The *focus* is where the light rays are brought together at a point and the focal length is the distance of the focus from the centre of the lens.

   A convex lens forms a small image of the object which is inverted (upside down) and on the opposite side of the lens. This is called a *real image* and can be seen on a screen. As an object in the distance is brought nearer to a convex lens the image moves away from the lens and becomes larger. The image formed by a concave lens is always upright, smaller than the object and it is a *virtual image*, that is it cannot be projected because although the rays of light appear to come to the observer from the image, they do not actually do so.

   Lenses can be found in all sorts of optical instruments including the camera and the telescope and also in the human eye (see SIGHT).

Leonids a meteor swarm whose ORBIT crosses the Earth's orbit annually. The peak of activity usually occurs around 17 November and there are occasionally spectacular displays.

leprosy an infectious disease that affects the skin, nerves and mucous membranes of the patient. The symptoms of leprosy include severe lesions of the skin and destruction of nerves, which can lead to disfigurements such as wrist-drop and claw-foot. Leprosy is caused by the airborne BACTERIUM, *Mycobacterium lepra*, and, fortunately, is not highly contagious as transmission requires direct contact with this bacterium. The likeliest source of infection, therefore, arises from the nasal secretions (swarming with bacteria) of patients and not from the popular misconception of touching the skin of an infected individual. Leprosy is curable, and the treatment, using sulphone drugs, has the beneficial side-effect of making the patient non-infectious even if he or she is not completely cured. Although the incidence of leprosy was once worldwide, it is now mostly confined to tropical and subtropical regions.

lepton see **elementary particle**.

lethal gene a gene that, if expressed, will cause the death of the individual. The fatal effect of the expressed gene usually occurs in the prenatal developmental stage of the individual, i.e. the embryonic stage for animals and the pupal stage for insects. Although most examples of lethal mutants fail to survive to adulthood, there is one well-researched genetic disorder, called Huntington's chorea, which does not usually affect the individual until middle age. Huntington's chorea is caused by a single dominant gene, and thus half the children of an affected parent will inherit the genetic disorder, although fortunately this disease is rare.

leucocyte or white blood cell is a large, colourless cell formed in the bone marrow and found in the blood of all normal vertebrates. It plays an important role in the IMMUNE SYSTEM of an individual. Leucocytes are also produced in the spleen, thymus and lymph nodes of the body and can be classified into the following three groups in order of decreasing content of leucocytes:

Group	%	Functions
Granulocyte	70	Helps combat bacterial and viral infection and may also be involved in allergies.
LYMPHOCYTE	25	Destroys any foreign bodies either directly (T-CELLS) or indirectly by producing antibodies (B-CELLS).
Monocyte	5	Ingests bacteria and foreign bodies by the mechanism called PHAGOCYTOSIS.

leucoplast a colourless object that contains reserves of starch and is found in some plant cells. If a leucoplast contains the pigment CHLOROPHYLL it may develop into a CHLOROPLAST.

leukaemia a cancerous disease in which there is an uncontrolled proliferation of white blood cells (LEUCOCYTES) in the bone marrow. The white blood cells fail to mature to adult cells and thus they cannot function as an important part of the defence system against infections. Although the definite cause of leukaemia is as yet unknown, there is growing suspicion that certain viruses may be involved and that perhaps there is an hereditary component. Unfortunately, leukaemia is not a curable disease, but there are methods effective in suppressing the reproduction of white blood cells—radiotherapy and, more commonly, chemotherapy. These methods bring the disease under control and thus help prolong the patient's life.

levée a bank or ridge running along the edges of a river and sloping away from the water. It is caused by deposition of coarse sands and silts when the river floods. If built up by man or by repeated flooding, it may permit the river level to rise above its flood plain with the risk of greater damage should further flooding occur.

lever a simple machine which at its simplest consists of a rigid beam which pivots at a point called the *fulcrum*. A load applied at one end can be balanced by an *effort* (a force) applied at the opposite end. There are three classes of lever depending upon the position of fulcrum, load and effort. The fulcrum may be between the effort and load (sometimes called a first-class lever); if the load is between the fulcrum and effort it is second class; and a third class lever has the effort between the fulcrum and load. Examples of these are pliers, a wheelbarrow and the shovel on a mechanical digger.

Although the work done by both effort and load must be equal, it is possible by moving a small effort through a large distance to move a very large load, albeit through a small distance.

librations the oscillation of a satellite e.g. the Moon, when viewed from Earth. It is due to a PARALLAX effect.

lichen a plant-like growth formed from two organisms which live together in a symbiotic relationship (see SYMBIOSIS). The two organisms involved are a FUNGUS and an ALGA. A lichen forms a distinct structure which is not similar to either partner on its own. Usually most of the plant body is made up of the fungus with the algal cells distributed within. The fungus protects the algal cells and the alga provides the fungus with food through PHOTOSYNTHESIS. A lichen is typically very slow growing and varies in size from a few millimetres to several metres across. It may form a thin flat crust, be leaf-like or upright and branching. Often, lichens are found in conditions that are too cold or exposed for other plants such as Arctic and mountainous regions. Reindeer moss and Iceland moss are lichens that provide an important food source in Arctic regions. Other lichens contain substances used in dyes, perfumes, cosmetics and medicines, as well as poisons.

ligand any MOLECULE or ATOM capable of forming a bond with another molecule (usually a metallic CATION) by donating an ELECTRON PAIR to form a complex ION. In biological terms, ligand refers to any molecule capable of binding with a specific ANTIBODY.

light is ELECTROMAGNETIC WAVES of a particular wavelength which are visible to the human eye. Objects can only be seen if light reflected from them or given off by them reaches the eye. Light is given out by hot objects and the hotter the object the nearer to the blue end of the spectrum is the light emitted. The Sun is our primary source of light and the light travels at almost 3×10^5 (300 000) km/s through space. The light waves are made up of packets or quanta (singular, quantum) of energy, called *photons*. Every photon can be considered to be a particle of light energy, the energy increasing as the wavelength shortens.

White light is actually made up of a range of colours. This can be shown by passing a very narrow beam of light through a prism. Because the individual colours have different wavelengths they are refracted by differing amounts and the resulting *dispersion* produces the characteristic *spectrum*.

A beam of light can often be seen, particularly if a torch is shone on a misty night or in a dark room where there is smoke, or if sunlight highlights dust particles. In each case the edge of the beam indicates that light travels in straight lines and because of this a shadow forms when an object is placed between a light source and a flat surface.

light-curve the graphical plot obtained from measuring the change in brightness of a VARIABLE STAR with time.

lightning the discharge of high-voltage electricity between a cloud and its base, and between the base of the cloud and the earth. One flash of lightning actually consists of several separate strokes that follow each other at intervals of fractions of a second (too fast for the human eye to detect) and occurs when the strength of the ELECTRIC FIELDS becomes great enough to overcome the RESISTANCE of the air. The discharge to the ground is followed by a return discharge to the cloud and this is the visible part of lightning. The clap of thunder that can be heard during thunderstorms is a result of the expansion of the air in between.

light reactions the biochemical processes that generate ATP, oxygen and other products during PHOTOSYNTHESIS, in the presence of light. The light-dependent reactions occur in the inner membranes of CHLOROPLASTS and require water and several forms of the pigment CHLOROPHYLL. Some of the products of the light reactions, including ATP, enter the CALVIN CYCLE, whereas the OXYGEN is released by the plant.

light wave *see* **electromagnetic waves**.

light year a measure of the distance travelled by light in one year, which is calculated in the region of 9.467×10^{12} kilometres or 9,460,700,000 km. Light travels at 299,792 kilometres per second and it takes 8 minutes for it to reach Earth from the Sun. The great distances between the various galaxies and heavenly bodies can thus be measured in light years. The nearest star to Earth (after the Sun), is more than four light years distant and is called *Proxima Centauri* (see also PARSEC).

lignite (or brown coal) a coal which in RANK falls between peat and bituminous coal. It shows little alteration, and the woody structure is often visible. It has a high moisture content.

limb the areas between hinges in folded rocks (see FOLD) In astronomy, the rim of a heavenly body that has a visible disk e.g. Sun or Moon.

limestone a SEDIMENTARY rock composed mainly of CALCITE with DOLOMITE and which may be organic, chemical or detrital in origin. There is a tremendous variety in the make up of limestones, which may comprise remains of marine organisms (corals, shells, etc); minute organic remains (as in CHALK); grains formed as spherically layered pellets (see OOLITE) in shallow marine waters and calcareous muds. Recent deposits of CALCIUM CARBONATE are found in shallow tropical seas. Modification after deposition is usually extensive and both the composition and structure may change due to compaction, recrystallization and replacement.

Limestone has many uses commercially including building stone, roadstone and aggregate, as a source material in the chemicals industry and in their natural state as AQUIFERS and petroleum RESERVOIR rocks.

limiting reactant any substance that limits the quantity of the product obtained during a chemical reaction. The limiting reactant can be identified using the chemical equation of a reaction as it will be the smallest quantity in comparison to the other reactants and products.

limonite a secondary 'mineral', and a hydrated iron oxide. It is usually amorphous (no crystalline shape) and as a weathering product of mineral deposits or iron in rocks can occur in considerable deposits. It is the major constituent of bog iron ores which form on lake floors due to bacterial action.

lineament a long structural or volcanic feature on the Earth's surface.

linear equation a mathematical term used to describe any equation containing two VARIABLES and of the general form:

$$y = mx + c$$

where x and y are the variables, m is the slope of the line and c is the intercept or the point where the curve crosses the y-axis. Linear equations are used extensively in engineering calculations.

line of sight velocity the velocity at which a heavenly body approaches or recedes from the Earth.

line squall when warm air runs over cold air it produces a squall line, often associated with a cold front, along which there are stormy conditions. It is characterized by sudden changes in temperature and pressure producing wind, heavy cloud and thunderstorms. The squalls may occur simultaneously, in a line across the country.

linkage the association between two or more GENES situated on the same CHROMOSOME. The genes tend to be inherited together, thus the parental gametes, which after FERTILIZATION eventually form the offspring, will not have undergone normal GENETIC RECOMBINATION to generate new combinations of genes. As the distance separating two genes decreases, the chance of these two genes becoming separated during CROSSING-OVER also decreases, thus increasing the chance that these genes are linked to be inherited as one segment rather than separate genes.

lipase any enzyme capable of breaking down fat to form FATTY ACIDS and GLYCEROL. Lipases function in alkaline conditions and are most abundant in the pancreatic secretions during digestion.

lipids a term that includes oils, fats, waxes and related products in living tissues. Lipids are insoluble in water but do dissolve in organic solvents. They form three groups: simple lipids, including fats, oils and waxes; compound lipids, which includes PHOSPHOLIPIDS; and derived lipids which includes STEROIDS. Their properties and functions are numerous and include energy storage, cell membrane components, some form vitamins and others occur in hormones.

lipoprotein any protein that has a FATTY ACID as a side chain. Lipoproteins are very important in certain biological processes as they function as a means of transport for essential molecules. For example, as CHOLESTEROL is extremely hydrophobic (that is, it does not mix with water), there would be no method of transporting it to its target body tissues. This problem is solved by low-density lipoprotein (LDL) surrounding the cholesterol molecule and forming a hydrophilic molecule (can mix with water) that can be transported by body fluids.

liquefaction of gases a gas may be turned into its liquid form by cooling below its critical temperature (the temperature above which a gas cannot be liquefied by PRESSURE alone). In addition, pressure may be required. For gases such as oxygen, helium and nitrogen, low temperatures are used. This enables bulk storage and transport of gases on a commercial scale.

liquid a fluid state of matter that has no definite shape and will acquire the shape of its containing vessel as it has little resistance to external forces. A liquid can be regarded as having more KINETIC ENERGY than a SOLID but less kinetic energy than a GAS. It is considered that the average kinetic energy will increase as the temperature of the liquid rises.

liquid crystal one of certain liquids that show regions of aligned molecules that are similar to crystals. The application of a current disrupts the molecules sufficiently to darken the liquid. This property is used in electronic equipment, for example, characters on the display window of a calculator.

lithification the processes which change unconsolidated (loose, uncemented) sediment into rock, including cementation of the grains.

lithology the description of a rock through its grain size, structure, mineral content (as far as possible) and general appearance.

lithosphere that layer of the Earth, above the ASTHENOSPHERE, which includes the crust and the top part of the mantle down to 80-120 kilometres in the oceans and around 150 kilometres in the continents. The base is gradational and varies in position depending upon the tectonic and volcanic activity of the region. The lithosphere contains the blocks that make up the process called PLATE TECTONICS.

litmus (see INDICATOR) a natural compound obtained from lichens which is used as an indicator, turning red to indicate acid conditions and blue for alkaline.

litre a unit of volume given the symbol l and equal to 1000 cubic centimetres, i.e. $1 l = 1000 cm^3 = 1 dm^3$. One gallon is approximately 4.5 litres.

liver the liver is a large and very important organ present in the body of vertebrate animals, just below the ribs in the region called the abdomen. It is composed of many groups of liver cells (called lobules) which are richly supplied with blood vessels. The liver plays a critical role in the regulation of many of the processes of METABOLISM. A vein called the hepatic portal vein carries the products of digestion to the liver. Here any extra glucose (sugar) which is not needed straight away is converted to a form in which it can be stored, known as glycogen. This is then available as a source of energy for muscles. PROTEINS are broken down in the liver and the excess building blocks, (amino acids), of which they are composed, are changed to ammonia and then to urea, a waste product which is excreted by the kidneys. Lipids, which are the products of the digestion of fat, are broken down in the liver and CHOLESTEROL, an essential part of cell membranes, some hormones and the nervous system, is produced. Bile, which is stored in the gall bladder and then passed to the intestine, is produced in the liver. In addition, poisonous substances (toxins) such as alcohol are broken down (detoxified) by liver cells. Important blood proteins are produced as are substances which are essential in blood clotting. Vitamin A is both produced and stored in the liver and it is a storage site for vitamins D, E and K. Iron is also stored and some hormones and damaged red blood cells are processed and removed in the liver.

loam a type of soil with sand, silt and clay in roughly equal proportions, often with some organic matter.

local gravitational constant the quantity given to the ACCELERATION of any object near to sea level at any point on earth. This acceleration is a result of gravity and is given the symbol g. At the North Pole, for example, $g = 9.8321$ ms^{-2}, and at the equator $g = 9.7801$ ms^{-2}.

local group the cluster of galaxies which includes the MILKY WAY. There are approximately 20-25 members within a radius of about 3 million light years. The largest member is the Andromeda galaxy.

locus (*plural* **loci**) the set of specific points that either satisfies or is determined by a certain mathematical condition. The locus can be thought of as tracing the path of a moving point relative to another fixed point. For example, a circle is a locus of a point that moves in such a way that the distance (radius of circle) between the moving point and the fixed point (centre of the circle) is constant. In biology, the locus is the name given to the region of a CHROMOSOME occupied by a particular GENE.

lode a vein or fissure in a rock containing mineral deposits.

loess a sediment formed from the aeolian transportation (i.e. carried by the wind) and deposition of mainly silt-sized particles of QUARTZ. It is well sorted, unstratified, highly porous and although it can form steep and even vertical slopes, is readily reworked. Loess is widespread geographically and although thicknesses of Chinese deposits exceeds 300 metres, it is normally just a few metres. Its origin has been hotly debated but is now accepted that the loess particles are produced by glacial grinding, frost cracking, hydration in desert regions and aeolian impact of sand grains (that is, due to the wind). The wind is the primary and essential agent in the process.

logarithm abbreviated to log, is a mathematical function first introduced to render multiplication and division with large numbers more simple. However, the advent of calculators and computers has reduced the former dependency on logs. The basic definition is that if a number x is expressed as a *power* (i.e. the number of times a quantity is multiplied by itself) of another number, y, that is $x = y^n$, then n is the logarithm of x to the base y, written as $\log_y x$.

There are two types of log in use; *common* (or *Briggs'*) and *Napierian* or *natural*. Common logs have base 10, $\log_{10} x$. Addition of the logs of two numbers gives their *product*, while subtraction of two numbers' logs is the means of division. Natural logs are to the base e where e is a constant with the value 2.71828. The two logs can be related by the function:

$$\log_e x = 2.303 \log_{10} x$$

lone pair a pair of electrons not shared by another atom but which can form CO-ORDINATION COMPOUNDS.

longitude the angular distance of a given point on the Earth's surface relative to the Greenwich meridian (a theoretical line that runs through the North and South Poles as well as Greenwich, England). Longitude is measured in degrees east and west of the Greenwich meridian (which is given an arbitrary value of 0°). By international agreement, the world is divided into 24 longitudinal zones, each 15° in width, starting and finishing at Greenwich (0°), to be able to relate the different times in places throughout the world. Thus, any place in the zone centred at 15° east of Greenwich is one hour ahead of Greenwich, whereas any place in the zone centred at 15° west is one hour behind Greenwich.

longitudinal wave the classification for a wave that is produced when the vibrations occur in the same direction as the direction of travel for that wave. The most well-known example of a longitudinal wave is a SOUND wave, which is propagated (moved forward) by displacement of air particles, causing areas of high density (called compressions) and areas of low density (called rarefactions). The WAVELENGTH of a longitudinal wave is the combined length of a single compression and rarefaction. For the FREQUENCY and speed of a longitudinal wave, see WAVE.

longshore drift the movement of material (sand and shingle) along the shore by a current parallel to the shore line (i.e. a longshore current). Longshore drift occurs in two zones : beach drift, at the upper limit of wave activity (i.e. where a wave breaks on the beach) and the breaker zone (where waves collapse in shallow water) where material in suspension is carried by currents.

loudspeaker is a device for turning an electric current into sound and is found in radios, televisions, and many other pieces of equipment that output sound. The commonest design is the *moving-coil loudspeaker*. This consists of a cylindrical magnet with a central south pole surrounded by a circular north pole producing a strong radial magnetic field. There is also a coil sandwiched between the two poles of the magnet which is free to move forwards and backwards and a stiff paper cone that is fastened to the coil. Because the wire of the coil is positioned at right angles to the magnetic field, when current flows through the coil, it moves.

When an alternating current passes through the coil it moves forwards and backwards and the paper cone vibrates resulting in sound waves.

low in meteorology, is an area of low pressure - see DEPRESSION.

lubricant any substance that, when applied between surfaces, will cause a decrease in FRICTION. Some common lubricants include oil and graphite dust. Without lubrication, surfaces grind against each other creating wear and this can cause damage in, or shorten the life of, an engine.

luminescence the emission of light by a living organism that is not a consequence of raising the body temperature of the organism. For luminescence to occur, the cells of the organism must contain the protein called luciferin, the enzyme luciferase, and the energy source for the reaction, ATP. This is sometimes called bioluminescence. Luminescence also occurs when a body gives out light due to a cause other than a high temperature. It is due to a temporary change in the electronic structure of an atom, and involves an electron taking in energy and moving to a higher orbit in the atom which is then re-emitted as light when the electron falls back to its original orbit. The energy required to promote the electron to a higher orbit may come from light (*photoluminescence*) or from collisions of the atoms with fast particles (*fluorescence*). When materials continue to give out light after the primary energy source has been removed, this is called *phosphorescence*.

This phenomenon of luminescence is put to use in the CATHODE RAY TUBE of TELEVISIONS.

luminosity the amount of light radiated by a star, expressed as a magnitude.

lungs are the sac-like organs which are used for RESPIRATION in air-breathing vertebrate animals. In mammals a pair of lungs is situated within the rib cage and each is made of a thin, moist membrane which is highly folded, and it is here that oxygen is taken in to the body and carbon dioxide is given up. The lungs are filled and emptied by the muscular movement of a sheet-like layer dividing the thorax from the abdomen, known as the diaphragm. This is accompanied (and the effect made greater) by outward and inward movement of the ribs which are controlled by other muscles.

In mammals such as man, air enters through the nose (and mouth) and passes into the windpipe (or trachea) which itself branches into two smaller tubes called bronchi (*singular* bronchus). Each bronchus goes to one lung and further divides into smaller, finer tubes known as bronchioles. Each bronchiole is surrounded by a tiny sac called an alveolus (*plural* alveoli) formed from one minute fold of the lung membrane. Carbon dioxide passes out from the capillaries across the alveoli into the lungs and oxygen passes into the blood in the opposite direction, this process being known as gaseous exchange. The numerous folds of the membrane forming the alveoli increase the surface area over which this is able to take place.

lymph a colourless, watery fluid that surrounds the body cells of vertebrates. It circulates in the LYMPHATIC SYSTEM but is moved by the action of muscles, as opposed to the contraction of the heart. Lymph consists of 95 per cent water but contains protein, sugar, salts, and LEUCOCYTES, and carries fat from the gut wall during digestion.

lymphatic system a network of tubules found in all parts of a vertebrate's body except, if present, the central nervous system. The tubules drain the body fluid, LYMPH, from the tissue spaces, and as they gradually unite to form larger vessels, the lymph vessels finally drain into two major vessels that empty into veins at the base of the neck. The system consists of vessels and nodes, where the fluid is filtered, bacteria and other foreign bodies are destroyed, and LYMPHOCYTES enter the lymphatic system. The lymphatic system is not only an essential part of the IMMUNE SYSTEM, but also carries excess protein and water to the blood and transports digested fats.

lymphocyte a cell that is produced in the bone marrow and is an essential component of the IMMUNE SYSTEM. Lymphocytes are a form of white blood cell (LEUCOCYTE), which collect in the lymph nodes of the lymphatic system and defend the body against foreign bodies and bacterial infections. Lymphocytes are also present in blood but in a smaller percentage.

lysis the destruction of cells, commonly BLOOD CELLS, by antibodies called **lysins**. More generally, the destruction of cells or tissues by pathological processes, e.g. autolysis, where cells are broken down by enzymes produced in the cells undergoing breakdown.

lysozyme an enzyme which is present in tears, nasal secretions and on the skin and has an antibacterial action. It breaks the cell wall of the bacteria leaving them open to destruction. Lysozyme also occurs in egg-white, and it was the first enzyme for which the three-dimensional structure was determined.

M

Mach number the speed of a body expressed as a ratio with the SPEED OF SOUND. Mach 1 is sonic speed, and thus below 1 is subsonic and above 1 is supersonic.

mackerel sky a cloud pattern made up of wavy CIRRO- or ALTOCUMULUS with holes, suggesting the markings of a mackerel.

macrophage a special type of cell that forms part of the IMMUNE SYSTEM of vertebrates. Macrophages are derived from MONOCYTES in the blood system and can move to infected or inflamed areas of the body using PSEUDOPODIA. In these areas, they surround, engulf and degrade broken cells and other debris, including microbes, by means of the process called PHAGOCYTOSIS.

Magellanic clouds two separate GALAXIES, detached from the Milky Way, which appear, from the southern hemisphere, as patches of light. The Large Magellanic Cloud is approximately 170,000 LIGHT YEARS away and the Small Magellanic Cloud about 210,000 light years away. They contain a few thousand million stars.

magma the fluid rock beneath the Earth's surface, which solidifies to form IGNEOUS ROCKS. During volcanic eruptions, the LAVA extruded at the Earth's surface is not necessarily the same in composition as the magma that arises to form lava, since the magma may have lost some of its gaseous elements and some solids of the magma may have crystallized.

magnetic bubble a portion of computer memory which consists of a small region in a material such as garnet (a silicate mineral), which is magnetized in one direction. Slices of this material placed on a SUBSTRATE (base material) produce a magnetic chip that under a magnetic field produces magnetic bubbles. Information can be stored on the chip, which may contain up to one million bubbles in 20 square millimetres, in BINARY form, through the presence or absence of a bubble in a specific location.

magnetic field the region of space in which a magnetic body exerts its force. Magnetic fields are produced by moving charged particles and represent a force with a definite direction. There is a magnetic field covering all of the earth's surface, which is believed to be a result of the iron-nickel core.

magnetic storm a disturbance of the Earth's magnetic field due to solar particles, after a solar FLARE.

magnetism is the effective force which originates within the Earth and which behaves as if there were a powerful magnet at the centre of the Earth, producing a magnetic field. The magnetic field has its north and south poles pointing approximately to the geographic north and south poles and a compass needle or freely swinging magnet will align itself along the line of the magnetic field. With the correct instrument it can also be seen that the magnetic field dips into the Earth, increasing towards the poles.

A *bar magnet* has a north and south pole, so named because the pole at that end pointing to the north is called a north-seeking pole, and similarly with the south pole. When dipped in a material that can be magnetized, such as iron filings, the metal grains align themselves along the magnetic field between the poles of the magnet. Some materials can be magnetized in the presence of a magnet, e.g. iron and steel. Iron does not retain its magnetism, but steel does. These are called *temporary* and *permanent magnets*. A more effective way to produce a magnet is to slide a steel bar into a solenoid (coil) through which current is passed and magnetism is *induced* in the steel (see ELECTROMAGNET).

In addition to iron, cobalt and nickel can also be magnetized strongly and these materials are called ferro-magnetic. Non-metals and other metals such as copper, seem to be unaffected by magnetism, but very strong magnets do show some effect.

The origin of magnetism is actually unknown, although it is attributed to the flow of electric current. On the electronic scale within magnetic materials it is thought that electrons act as minute magnets (because electrons carry a charge) as they spin around their nuclei in atoms. In some elements, this electron spin is cancelled out but in others it is not and each atom or molecule acts as a magnet contributing to the overall magnetic nature of the material.

magnetosphere the space around the Earth and other planets with a magnetic field in which charged particles are affected by the magnetic field of that planet rather than of the Sun. The extent of the space is greater on the side of the planet away from the Sun by a factor of about 4. The boundary of the magnetosphere is well-defined but is altered by solar activity.

magnitude the absolute value or length of a physical or mathematical quantity.

In astronomy, a relative measure of the apparent brightness of stars. Based upon a logarithmic scale, a difference of one in magnitude is actually a ratio of 2.51, thus a star of a particular magnitude is 2.51 times brighter than a star of a magnitude which is one less. The lower the numerical value of the magnitude, the brighter is the star.

The *absolute* magnitude is the magnitude a star would have if it were placed 10 PARSECS away from Earth.

main sequence in astronomy a band within the HERTZSPRUNG-RUSSELL DIAGRAM in which most stars lie and range from high temperature and LUMINOSITY to low temperature and luminosity (depending mainly upon mass). During much of its life, while converting HYDROGEN to HELIUM a star lies in the main sequence and then moves out as the hydrogen is consumed, becoming a RED GIANT initially before evolving into other forms (SEE STELLAR EVOLUTION).

malaria an infectious disease caused by certain parasites in the blood of the victim. Malaria is transmitted by an infected mosquito biting a human, thus injecting the parasite from the salivary gland of the mosquito into the bloodstream of the human. After the parasites have developed in the victim's liver, they are released into the bloodstream and attack red blood cells. Early symptoms of malaria include headaches, body aches and chills. As the disease progresses, malarial attacks are frequent and cause sickness, dizziness and sometimes delirium in the victim; the attacks seem to coincide with the bursting of infected red blood cells. Fortunately, there are many highly effective drugs for treating malaria.

malleability the property of metals and alloys that enables them to be changed in shape by hammering or rolling (or similar processes) into thin sheets.

manganese nodules small (on average 2 cm) nodules containing manganese oxides, iron oxides with nickel, copper, zinc, cobalt and traces of other elementals. The nodules, found on sea and lake floors are particularly associated with the Pacific Ocean. The nodules show concentric layering and probably formed from the alteration of organic matter and other substances on the sediment of the sea floor. The concentration of valuable metals means the nodules are a source of ore but their recovery is far from straightforward as they are found at depths of up to 4500 metres and across vast areas.

manometer an instrument to measure pressure, often in the form of a U-tube containing mercury.

mantle that layer of the Earth between the crust and core which reaches thicknesses of about 2500 kilometres. The boundary is marked by the MOHOROVICIC DISCONTINUITY. Its composition approximates to that of garnet peridotite (peridotite is an ULTRABASIC IGNEOUS rock with the minerals olivine, pyroxene and chromite, $FeCr_2O_4$).

marble a limestone which has been metamorphosed (see METAMORPHIC ROCKS) and recrystallized due to the action of heat. The alteration renders marble hard enough to take a polish and it has been and still is used extensively for building, ornamental and decorative work.

maria the term, from Latin, for the 'seas' on the surface of the MOON. The name was coined before modern study found them to be dry, but their origin is not established although it is thought they date from 3300 million years ago. The tendency has been to drop the use of the Latin and use the English equivalent e.g., Sea of Tranquility.

Mars is the fourth planet in the SOLAR SYSTEM and the one nearest to the Earth. Its orbit lies between that of the Earth and Jupiter, and it is about half the size of Earth. It has a thin atmosphere, exerting a pressure less than one hundredth of that of the Earth. It also has a small mass about one tenth of that of the Earth so that a person on Mars would weigh about 60% less than Earth weight. Mars is often called the *Red Planet* as it has a dusty, reddish surface strewn with rocks. It is much colder than the Earth with an atmosphere mainly of carbon dioxide which is frozen at the two poles. The polar ice caps melt and re-form as the seasons change. Minimum surface temperatures are in the region of -100°C with the maximum only about -30°C. A year on Mars lasts for 687 Earth days and the length of one day is almost the same, 24 hours and 37 minutes. The diameter of Mars is 6794 km and the crust in the northern part of the planet is composed of basalt (volcanic rock). There are many extinct volcanoes, canyons and impact craters and evidence of water erosion at some stage in the planet's history. The mountains are much higher, and the valleys deeper than those which exist on Earth, hence there must have been violent movements of the crust during the past. The deepest valley called *Valles Marineris*, is 4000 km long, 75 km wide and up to 7 km deep. The highest mountain, *Olympus Mons*, rises 23 km from the surface of Mars, and is three times taller than Mount Everest.

There are two small, irregularly shaped SATELLITES or moons orbiting Mars, called *Phobos* and *Deimos*. Phobos, the largest, is only 27 km from one end to the other and just 6000 km above the planet's surface. Its orbit is a gradually descending spiral and it is estimated that in 40 million years' time it will collide with Mars. The *Viking* space probe landed on Mars in 1976 and many valuable photographs were taken and soil samples obtained. There had been speculation about the possibility of the existence of life on Mars for many years but the space probe failed to discover any evidence for this. Mars is currently being studied by means of the *Surveyor* spacecraft.

maser (abbreviation for Microwave Amplification by Stimulated Emission of Radiation) a microwave amplifier/oscillator working in a similar way to the laser.

Maser oscillations produce a narrow beam of monochromatic (very narrow frequency band) radiation.

mass the measure of the quantity of matter that a substance possesses. Mass is measured in grams (g) or kilograms (kg).

mass number (A) the total number of PROTONS and NEUTRONS in the nucleus of any atom. The mass number therefore approximates to the RELATIVE ATOMIC MASS of an atom.

mass spectrometer a machine used to detect the various types of ISOTOPES found in an element. The mass spectrometer bombards molecules with high-energy electrons, creating smaller positive IONS and neutral fragments. These positive ions are then deflected, using a MAGNETIC FIELD and each is deflected differently and separated according to their varying MASS. The ions finally pass through a slit to the ion collector, and each peak of the printed chart corresponds to a particular ion and its mass. The mass spectrometer also provides information on the relative amount of each isotope present as well as the exact mass of the various isotopes.

mast cell a large cell carried in the blood that has a fast-acting role in the IMMUNE SYSTEM. In allergic reactions, mast cells will be triggered to release histamine when the IMMUNOGLOBIN, IgE, has already attached itself to the foreign particle that has entered the body.

matrix (*plural* **matrices**) an array of elements, that is numbers or algebraic symbols, set out in rows and columns. It may be a square or rectangular arrangement and the order of a matrix refers to the number of rows and columns, e.g.:

$$A = \quad (4 \ 1 \ 6) \qquad B = \begin{matrix} 2 & 6 \\ 1 & 4 \end{matrix}$$

Matrix A has one row, three columns, and matrix B has two rows, two columns. Only matrices of the same order can be added or subtracted. Matrices are only compatible for multiplication if the number of columns of the first equals the number of rows of the second. To multiply, therefore, the row of one matrix is multiplied by the column of the other matrix, and the products are added. Matrices are useful for condensing information and solving complex LINEAR EQUATIONS.

matter any substance that occupies space and has MASS: the material of which the universe is made. It normally exists in the three states of gas, liquid or solid.

Plasma is considered to be a fourth state but it only exists at very high temperatures e.g., in stars.

mean daily motion the angle a celestial body moves through in one day, assuming uniform orbital motion.

meander the side to side wandering of a stream/river channel, which is best developed in river deposits on the flood plain. The origin of meanders is not established but it involves factors such as the original stream course and the natural physical properties of water when flowing over sediment or rock. Also, a river will take the course which requires the least energy to follow. Once established, meanders create erosion on the outer banks of the river where flow is faster and deposition on the inner banks where flow is checked such that water cannot carry so much sediment. The curved river form can then become more exaggerated and 'migrate' downstream and a variety of features may be developed including OXBOW LAKES.

mean solar day is defined as the average value of the interval between successive returns of the SUN to the MERIDIAN.

megabyte (MB) see **byte**.

megaparsec a unit for defining distance of objects outside the galaxy, equal to 10^6 PARSEC (or 3.26×10^6 light years).

meiosis a division of the chromosomes that produces the germ cells (GAMETES in animals and some plants, sexual spores in fungi). Meiosis involves the same stages as MITOSIS, but each stage occurs twice, and, as a consequence, four HAPLOID cells are produced from one DIPLOID cell i.e., four gametes are produced with half the number of chromosomes in the parent cell. Meiosis is an extremely important aspect of sexual reproduction, as the production of haploid cells ensures that during FERTILIZATION the chromosomal number is constant for every generation. It also gives rise to genetic variation in the daughter haploid cells by the rearrangement and GENETIC RECOMBINATION if genetic variability already exists in the parent diploid cell.

melanin a dark pigment responsible for the colouring of the skin and hair of many animals, including humans. Differences in skin colour are due to variations in the distribution of melanin in the skin and not to differences in the number of cells, (melanocytes) that produce the pigment.

melanism when some individuals in a population are black due to overproduction of the pigment MELANIN. It is particularly well noted with insects found in some industrial areas where, due to the increase in industrial pollution, a species evolved to compensate. In this case of *industrial melanism*, the peppered moth was found to be much darker in polluted areas compared to the non-polluted parts, thus making it less visible to a predator.

melting point the temperature at which a substance is in a state of equilibrium between the solid and liquid states, e.g. ice/water. At 1 atmosphere pressure, the melting point of a pure substance is a constant and is the same as the freezing point of that substance. At constant pressure, the melting point of a substance is lowered if it contains impurities (the reason for putting salt on ice to make it melt). Although an increase in pressure lowers the melting point of ice, in most substances increased pressure will raise their melting points.

membrane a membrane is a thin sheet of tissue widely found in living organisms. It covers, lines or joins cells, organelles (small organs) , organs and tissues and consists of a double layer of lipids (fats) in which protein molecules are suspended. Water and fat soluble substances are able to pass across a membrane but sugars cannot. Other substances or ions are actively carried across a membrane by a complex system known as *active transport.*

membrane potential the difference in electric potential between the inside and the outside of the plasma membrane of all animal cells. The inside of the cell is more negative than the outside and membrane potentials exist due to the different ionic concentrations in fluids within and outwith the cell. It is also due to the selective permeability of the plasma membrane to specific ions (notably potassium $[K^+]$, sodium $[Na^+]$ and chloride $[Cl^-]$ ions). When measured using a microelectrode, the resting potential, that is when the cell is not sending an electrical signal, of most muscle and nerve cells is -60mV on the inside of the plasma membrane.

Mendel's laws of genetics laws of heredity deduced by Gregor Mendel (1822–1884), an Austrian monk, who discovered certain basic rules following experiments he carried out with generations of pea plants. Mendel discovered that a trait, such as flower colour or plant height, had two factors (hereditary units) and that these factors do not blend but can be either dominant or recessive. Without knowing about genes or cell division, he developed the following laws of particular inheritance:

First law—each hereditary factor is due to two factors. It is now recognized that ALLELES ('factors') are present on HOMOLOGOUS CHROMOSOMES, which separate during MEIOSIS.

Second law—factors for different traits undergo INDEPENDENT ASSORTMENT. During gamete formation, the segregation of one gene pair is independent of any other gene pair unless the genes are linked (see LINKAGE).

meniscus the effect of surface tension in a liquid producing a definite surface.

Where this meets a solid e.g., water in a glass tube, this produces a rising of the water's surface up the tube. The reverse effect is seen with mercury because the cohesion (binding together) between mercury molecules is stronger than the adhesion between mercury and glass.

Mercury the smallest and first planet of the SOLAR SYSTEM and nearest to the SUN. It has no atmosphere, and so during the day the surface temperature reaches 425°C (enough to melt lead), but at night the heat all escapes and it becomes intensely cold -170°C. A day on Mercury lasts for 59 Earth days but the planet travels its orbit so fast that a year is only 88 days. Mercury is a very dense planet for, although it is only slightly bigger than the Moon, it has an enormous mass which is almost the same as that of the Earth. It is thought that this is accounted for by having a huge metallic core. Little was known about the surface of the planet until it was visited by the *Mariner 10* space craft which passed to within 800 km of Mercury in 1974. It revealed that Mercury has a wrinkled surface with thousands of craters which have been caused by the impact of meteors and other larger space bodies. The largest crater, 1300 km across and known as the *Caloris Basin*, must have been caused by the collision of an enormous space body. Mercury has very little gravity and an elliptical orbit, which takes it to within 46 million kilometres of the Sun at its nearest point and 70 million kilometres when farthest away.

mercury a silvery-white liquid metal element (and the only liquid metal element at room temperature) which occurs naturally as the ore cinnabar (HgS - mercuric sulphide). As an alloy with most metals it forms AMALGAMS. It is used extensively in thermometers, barometers and other scientific apparatus, and in the manufacture of batteries, drugs, chemicals including pesticides, etc. Mercury and many of its compounds are toxic.

mercury barometer an instrument for measuring atmospheric pressure denoted by a column of mercury (in a tube) which exerts an equal pressure.

meridian the great circle cutting the CELESTIAL SPHERE at its poles and which cuts the observer's horizon at the north and south points.

meson unstable particles belonging to the group called HADRONS. They are found in COSMIC RAYS and consist of a QUARK and its antiquark. Their mass is between that of an electron and a nucleon (a PROTON or NEUTRON) and there are positive, negative and neutral varieties.

mesophyll the internal tissues of a leaf that are between the upper and lower epidermal layers. Mesophyll tissue contains CHLOROPLASTS, which are concentrated at a site that allows maximum absorption of light for PHOTOSYNTHESIS.

messenger RNA (mRNA) a single-stranded RNA (ribonucleic acid) molecule that has the important role of copying genetic information from DNA in the nucleus and carrying this information in the form of a sequence of bases to special sites (RIBOSOMES) in the cell, where the specified protein that was encoded in the nuclear DNA is synthesized.

Messier catalogue a list compiled by the French comet hunter Charles Messier in 1770 which contains 108 galaxies, nebulae and star clusters. The names assigned to objects are frequently still used in astronomy.

metabolism is the name given to all the chemical and physical processes which occur in living organisms. These are of two kinds, *catabolic* (or breaking down) as in the digestion of food, and *anabolic* (or building up) as in the production of more complicated MOLECULES from simple ones. All these processes require energy and ENZYMES in order to take place. Plants trap energy from the sun during PHOTOSYNTHESIS and animals gain energy from the consumption of food. The metabolic rate is the speed at which food is used or broken down to produce energy, and this varies greatly between different species of animals. In people, children have a higher metabolic rate than adults and more energy is required by someone during hard work than by someone who is at rest.

metal a substance that has a "metallic" lustre or sheen and is generally ductile, malleable, dense and a good CONDUCTOR of electricity and heat. Elements with these properties are generally electropositive, i.e., give up electrons, becoming positively charged (e.g. Na$^+$) when combining with other atoms, molecules or groups. When combining with water, BASES result (e.g. NaOH, sodium hydroxide), and their chlorides (e.g. NaCl, sodium chloride) are stable towards water. Not all elements normally considered metals show all these properties. Elements with characteristics of both metals and non-metals are termed METALLOIDS. There are about eighty metals most of which occur as compounds in ores.

metal fatigue see **fatigue of metals**.

metalloid an ELEMENT which exhibits some properties associated with metals and some associated with non-metals. Metalloids also exhibit AMPHOTERISM.

metallurgy the scientific study of metals and their alloys, including extraction from their ORES, and processing for use.

metamorphic rocks one of the three main ROCK types, are formed by the alteration or recrystallization of existing rocks by the application of heat, pressure, change in volatiles (gases and liquids), or a combination of these factors.

meteor

There are several categories of *metamorphism* based upon the conditions of origin: *regional*—high pressure and temperature as found in *orogenic* (mountain-building) areas; *contact*—where the rocks are adjacent to an igneous body and have been altered by the heat (with little or no pressure); *dynamic*—very high, confined pressure with some heat, as generated in an area of faulting or thrusting, i.e. where rock masses slide against each other; *burial*—which involves high pressure and low temperature as found, for example, at great depth in sequences of sedimentary rocks.

The key feature of all metamorphic rocks is that the existing *assemblage* (group) of minerals is changed by the pressure and/or heat and the presence of fluids or other volatiles. New minerals grow that are characteristic of the new conditions. Some typical metamorphic rocks are schist, slate, gneiss, marble, quartzite and hornfels. Depending upon the type of metamorphism, there are systems of classification into *zones* or *grades* where specific minerals appear in response to increasing pressure and/or temperature.

metamorphosis the period of change in form of an organism from the larval to the adult state. Common examples include a tadpole changing into a frog and a caterpillar into a butterfly. The changes are controlled by hormones.

metaphase a stage of cell division in MITOSIS or MEIOSIS. During metaphase, the chromosomes are organized and attached to the equator of the spindle by their CENTROMERES. Metaphase occurs only once in mitosis but twice in meiosis.

metasomatism the introduction of chemical constituents in a gaseous or liquid phase into a rock (or their removal from it) thus altering the overall composition of the rock. Many occur with metamorphic processes, and certain minerals may be completely transformed.

metastable a term applied to a group of minerals created under high temperature and/or pressure, or a supersaturated solution (see SUPERSATURATION) which appears to be stable but, in fact, will react or change if disturbed. The state is due to the system being very slow in reaching equilibrium. A PHASE is termed metastable if it co-exists with another phase which is stable. There are many minerals and whole rock assemblages e.g. diamond, and a high pressured temperature METAMORPHIC rock respectively that are metastable at the surface. Such assemblages are temporarily in equilibrium.

metastasis the process in which malignant cancerous cells spread from the affected tissue to create secondary areas of growth in other tissues of the body.

meteor or shooting star, is a small body from the SOLAR SYSTEM which burns up on entering the Earth's atmosphere. The size varies from kilograms down to a fraction of a gram. Travelling at about 15-20 kilometres per second, friction in the atmosphere vaporizes small particles creating light.

meteoric water water originating in the atmosphere e.g. as rain and snow. (See *also* JUVENILE).

meteorite *see* **asteroid**.

meteorology the study of the processes and conditions (e.g. pressure, wind speed, temperature) in the earth's ATMOSPHERE. The resulting data enables predictions to be made as to likely future weather patterns.

meteor stream meteor shows caused by dust streams intersecting the Earth and its atmosphere. The streams of dust orbit the Sun.

methane the first member of the series of ALKANES. It is a colourless, odourless gas with the chemical formula CH_4. Methane is the main constituent of coal gas and is a byproduct of any decaying vegetable matter. It is flammable and is used in industry as a source of hydrogen. Methane also occurs in significant quantities in the atmospheres of Neptune, Uranus, Saturn and Jupiter.

methyl (or Me) the group CH_3.

methylation the process of adding a METHYL group to a compound. In biology it is the addition of a methyl group to a NUCLEIC ACID base, e.g. ADENINE or CYTOSINE.

metrology the science of measurement.

mica a group of aluminium, potassium SILICATES with magnesium and iron in dark varieties, and sodium, lithium or titanium in other lighter coloured forms. HYDROXYL (OH) is always present, often partly replaced by fluorine. Micas have a sheet-like structure and a perfect CLEAVAGE parallel to the sheets (which are made up of Si_4O_{10} groups). The group comprises several members including BIOTITE, muscovite or white mica (potassium variety) lepidolite (lithium potassium), phlogopite (magnesium), zinnwaldite (lithium) and glauconite (iron and extra aluminium). Micas, especially biotite and muscovite occur in many rock types, especially igneous and metamorphic rocks. Muscovite and phlogopite are particularly important in the electrical industry as insulators.

micrometer any instrument used for the accurate measurement of minute objects, distances or angles.

micrometre the unit of length that equals 10^{-6} of a metre and has the symbol μm.

microtubule a long, hollow fibre of PROTEIN that is found in all higher plant and animal cells. Microtubules have different functions in different cells, e.g. they form the spindle during MITOSIS, give strength and rigidity to the tentacles of some unicellular organisms, and they are also found in parts of nerve cells.

microwave the part of the electromagnetic spectrum (see ELECTROMAGNETIC WAVES) with a wavelength range of approximately 10^{-3} to 10 metres and a frequency range of 10^{11} to 10^{7}Hz. When absorbed, microwaves produce large amounts of heat, a useful property for the economical and quick cooking of food. The waves are generated by a unit called the magnetron. Microwaves are easily deflected, and as they have a shorter wavelength range than radiowaves they are more suitable for use in RADAR systems as they can detect smaller objects. They are also used for communications via satellites.

microwave background a weak radio signal discovered in 1963 which is thought to be the remains of the big-bang with which the UNIVERSE began. It occurs throughout space at an almost identical intensity.

mid-oceanic ridge (oceanic ridge, or simply ridge) throughout the world's oceans are long, linear volcanic ridges, which are in effect submarine mountain chains, generally positioned centrally. The ridges are sites where new oceanic crust is created through the spreading of the plates (a spreading axis) and outpouring of basalt. The mid-oceanic ridges are sites of shallow earthquakes and they are often cut by TRANSFORM FAULTS which create an offset appearance to the ridge and those faults linking the ridge are also the focus of earthquakes (see SEA-FLOOR SPREADING).

migration the seasonal movement of animals especially birds, fish and some mammals (e.g. porpoises). Climatic conditions usually trigger off migration where perhaps lower temperatures indicate less available food. Some animals, particularly birds, travel vast distances, e.g. golden plovers, fly 8000 miles from the Arctic to South America.

Milankovitch theory the theory that periodic changes in the amount of solar radiation received by the Earth led to a cycle of changes in climate and led to ice ages. It was postulated that the variations in radiation were due to, amongst other things, the eccentricity (that is, not uniform) of the Earth's orbit.

Milky Way see galaxy.

mimicry a resemblance of one species to another species which has evolved as species have adapted. Mimicry occurs in both the animal and plant kingdoms but is predominantly found in insects. The main types of mimicry are:

(1) Batesian mimicry, named after the British naturalist H. W. Bates (1825–92)—where one harmless species mimics the appearance of another, usually poisonous, species. A good example of this is the non-poisonous viceroy butterfly mimicking the orange and black colour of the poisonous

monarch butterfly. The mimic benefits as, although harmless, any predator learns to avoid it as well as the poisonous species.

(2) Mullerian mimicry, named after the German zoologist J. F. T. Müller (1821–97)—where different species, which are either poisonous or just distasteful to the predator, have evolved to resemble each other. This resemblance ensures that the predator avoids any similar-looking species.

mineral a substance, usually inorganic, with a definite and characteristic chemical composition and usually with a crystalline structure and certain physical properties including: hardness (see MINERALOGY) lustre (i.e. the way its surface reflects light), CLEAVAGE, colour, fracture and relative density. Some elements occur as minerals in their own right e.g. gold, diamond (carbon) but the vast majority of minerals are compounds of several elements, with oxygen being the commonest element, present as oxides in many cases. There are approximately two thousand minerals but the most common rocks contain individual minerals such as quartz and calcite, or minerals from about five or six groups including the clay minerals, PYROXENES, AMPHIBOLES, FELDSPARS and the MICAS. These are all silicate minerals, which form the most abundant of the rock forming minerals.

mineralogy the study of any chemical element or compound extracted from the earth. Mineralogy examines the mode of formation and physico-chemical properties of minerals. These are generally solid or crystalline and can be classified according to their chemical constitution, i.e. molecular, metallic or ionic composition or crystallography. Another method of classifying minerals is according to their comparative hardness, and the Mohs' scale arranges them in relative order from the softest, talc (no. 1) to diamond (no. 10). Each mineral will scratch a mineral lower on the scale. The full list is: 1—talc; 2—gypsum; 3—calcite; 4—fluorite; 5—apatite; 6—orthoclase; 7—quartz; 8—topaz; 9—corundum; and 10—diamond.

mirage a visual phenomenon due to the REFLECTION and REFRACTION of light. A mirage is seen wherever there is calm air with varying temperatures near the earth's surface. A common mirage in the desert is caused by the refraction of a downward light ray from the sky, so that it seems to come from the sand and, to any onlooker, it would appear that the sky is reflected in a pool of water. As the inverted image of a distant tree is also usually formed, the overall effect resembles a tree being reflected in the surrounding water.

mist water droplets in suspension which reduce visibility to not less than 1 km (see also FOG).

mitochondrion (*plural* **mitochondria**) a type of rod-shaped organelle found in the cytoplasm (cell contents except for the nucleus) of EUCARYOTIC cells which is surrounded by a double membrane. Mitochondria have been called the "power houses" of the cell as they are very important in the generation of energy in a form called ATP. ATP production is the end result of cellular respiration and provides energy for all metabolic processes. As mitochondria are the sites where this takes place, they are especially abundant in cells which require lots of energy such as those of muscles. Mitochondria contain a form of DNA, structures called ribosomes in which proteins are manufactured with numerous enzymes, each specific to a particular metabolic process.

mitosis the process by which a NUCLEUS divides to produce two identical daughter nuclei with the same number of CHROMOSOMES as the parent nucleus. Mitosis occurs in several phases:

(1) Prophase—the condensed chromosomes become visible, and it is apparent that each chromosome consists of two CHROMATIDS joined by a CENTROMERE.

(2) METAPHASE—the nuclear membrane disappears, a spindle forms, and each chromosome becomes attached by its centromere to the equator of the spindle fibres.

(3) Anaphase—each centromere splits, and one chromatid from each pair moves to opposite poles of the spindle.

(4) TELOPHASE—a nuclear membrane forms around each of the group of chromatids (now regarded as chromosomes), and the cytoplasm divides to produce two daughter cells. The stage before and after mitosis is called interphase, and the chromosomes are invisible during this phase as they have decondensed. This is an extremely important part of a cell's cycle as DNA replicates and required proteins are synthesized during interphase.

Möbius strip a ribbon of paper where one end has been twisted through 180 degrees before joining to the other end. The result is a single continuous surface containing a continuous curve. This configuration can be used in machines driven by belts, because in using both sides of the belt, wear is reduced.

moderator a substance used in a nuclear reactor to slow down fast NEUTRONS generated by NUCLEAR FISSION. The substances contain a light element e.g. DEUTERIUM in heavy water, which absorbs some energy upon impact with the neutron, but avoids capturing the particle. These slower neutrons are then more likely to participate in the ongoing fission process.

modulus the measure of the value quantity regardless of its sign. The modulus of a REAL NUMBER $|x|$ always gives a positive value, e.g. $|-6| = 6$.

Mohorovicic discontinuity (often called simply the Moho) a major seismic discontinuity (a break in rock properties at depth) discovered in 1909 from a study of the seismograms of the Yugoslav earthquake which occurred that year. The feature was highlighted by double sets of the two types of seismic waves, interpreted as the direct and refracted versions caused one by the crust and the second by the mantle. Although this 'boundary' was at first considered to be quite a distinct feature, it is now known to be more complex.

Mohs' scale see **mineralogy**.

molality the concentration of a solution expressed in MOLES of solute per one kilogram of solvent. Molality has symbol m and units of mol kg^{-1}.

molarity the number of MOLES of a substance dissolved in one litre of solution. Molarity has symbol c and units mol $litres^{-1}$.

mole the amount of substance that contains the same number of ELEMENTARY PARTICLES as there are in 12 grams of carbon. The number of atoms, ions or electrons in one mole is called AVOGADRO'S CONSTANT and equals 6.023×10^{23} mol^{-1}.

molecular biology the study of the structure and the function of large molecules in living cells and the biology of cells and organisms in molecular terms. This applies especially to the structure of PROTEINS and the nucleic acids DNA and RNA.

molecular electronics the use of molecular materials in electronics and optoelectronics. Presently liquid crystalline materials (LIQUID CRYSTALS) are used in displays, organic semiconductors in xerography, and other organic materials are used in devices where the macroscopic properties of the materials are involved. The long term aim of this mixed discipline is to devise molecules or small aggregates of molecules that can exceed the performance of silicon integrated circuits.

molecular formula the chemical formula that indicates both the number and type of any atom present in a molecular substance. For example, the molecular formula for the alcohol ETHANOL is C_2H_6O and indicates that the molecule consists of two carbon atoms, six hydrogen atoms and one oxygen atom.

molecularity the number of particles that are involved in a single step of a reaction mechanism. One step of a mechanism may be termed unimolecular, bimolecular or termolecular, depending on whether there are one, two or three reacting particles. The reacting particles can be ATOMS, IONS or MOLECULES.

molecular sieve a compound with a framework structure, usually a synthetic ZEOLITE, which can be used for the specific absorption of water, gases, etc. In effect, the framework creates a cage to trap the appropriate molecules and the size of the trap can be modified.

molecular weight the total of the atomic weights of all the atoms present in a molecule.

molecule the smallest chemical unit of an element or compound that can exist independently. Any molecule consists of ATOMS bonded together in a fixed ratio, e.g. an oxygen molecule (O_2) has two oxygen atoms bonded together and a carbon dioxide molecule (CO_2) has two oxygen atoms bonded to one carbon atom. Molecules may contain thousands of atoms.

mole fraction the ratio of the number of MOLES of a substance to the total moles present in a mixture.

moment the moment of a FORCE (or VECTOR) about a point is a measure of the turning effect of the force and is the force multiplied by the perpendicular distance from the point. The tendency of the force to produce a twist about any given point is measured by the moment, *about that point*. If the force is in newtons and the distance in metres, the moment is measured in newton metres (Nm).

momentum is the property of an object defined as the product of its velocity and mass and it is measured in kgm/s. Momentum is related to force as follows:

force = rate of change of momentum.

Changes in momentum occur mainly due to the interaction between two bodies. During any interaction the total momentum of the bodies involved remains the same, providing no external force, such as FRICTION, is acting. Newton's second law of motion states that the rate of change of momentum of a body is directly proportional to the force acting and occurs in the direction in which the force ends.

If a body is rotating around an axis then it has *angular momentum* which is the product of its momentum and its perpendicular distance from the fixed axis.

Momentum can be seen around us all the time, but one of the most obvious examples is the collision of balls on the snooker table.

monobasic acid an ACID that contains only one replaceable hydrogen atom per molecule. A monobasic acid will produce a normal SALT during a reaction with a suitable metal.

monoclonal antibody a particular ANTIBODY produced by a cell or cells derived from a single parent cell, i.e. a clone (with each cell being monoclonal). Antibodies thus produced are identical and have specific AMINO ACID sequences. Monoclonal antibodies are used in identifying particular ANTIGENS within a mixture, for example, in identifying blood groups. They are also used to produce highly specific vaccines.

monocotyledon the subclass of flowering plants that have a single seed leaf (cotyledon). Any flowering plant that contains its seeds within an ovary (sometimes forming a fruit) belongs to one of two subclasses, either monocotyledon or dicotyledon (which has two seed leaves). Monocotyledons also differ from dicotyledons in that they have narrow, parallel-veined leaves, their vascular bundles are scattered throughout the stem, their root system is usually fibrous, and their flower parts are arranged in threes or multiples of threes. Most monocotyledons are small plants such as tulips, grasses or lilies.

monocyte a large phagocytic cell (see PHAGOCYTOSIS), capable of motion, which is present in blood. Monocytes originate in the bone marrow and, after a short residence in the blood, move into the tissues to become MACROPHAGES.

monomer a simple molecule that is the basic unit of POLYMERS. Most monomers contain carbon-carbon double bonds but just have single bonds after they have undergone ADDITION POLYMERIZATION to form the long-chained polymer. ETHENE is a type of monomer that reacts with other ethene molecules under high pressure and temperatures to form the polymer, polyethene (polythene).

monosaccharides (see also SACCHARIDES) sugars with the general formula $C_nH_{2n}O_n$ (where n = 5 or 6). They are grouped into hexoses ($C_6H_{12}O_6$) or pentoses ($C_5H_{10}O_5$) on the number of CARBON atoms present and are either aldoses which contain the CHO group or ketoses containing the CO group.

monsoon in general terms, winds which blow in opposite directions during different seasons of the year, with features that are associated with widespread temperature changes over land and water in the subtropics. Monsoon winds are essentially similar in origin to LAND AND SEA BREEZES but occur on a very much larger scale geographically and temporally. The Indian subcontinent is subjected to a rainy season in the south westerly monsoon. Other areas affected by monsoons are Asia (East and South East), parts of the West African coast and North Australia.

monozygotic twins see **identical twins**.

Moon the Earth's one satellite, which orbits around the Earth at an average distance of 384,000 kilometres (238,600 miles). It has no atmosphere, water or magnetic field, and surface temperatures reach extremes of 127°C (261°F) and -173°C (-279°F). It takes nearly 28 days to complete its orbit around the Earth, and always presents the same face towards Earth. As it orbits around the Earth each 28 days, the Moon passes through *phases* from new to first quarter to full to last quarter and back to new again. One half of the moon is always in sunlight and the phases depend upon the amount of the lit half which can be seen from

Earth. A *new moon* occurs when the Earth, Moon and Sun are approximately in line with one another, and none of the lit half can be seen at all; it appears that there is no Moon. About a week later a small *sliver* of the lit half can be seen from Earth and this grows throughout the month until the whole of the lit half is visible at *full moon*. In the second half of the cycle, the amount of the lit half of the Moon which can be seen from Earth gradually declines once more. When the Moon is apparently growing in size, it is called *waxing* and when it is declining it is called *waning*.

The diameter of the Moon is 3476 kilometres and its mass is 0.0123 of that of the Earth. Its density is 0.61 of that of the Earth and it has a thick crust (up to 125 kilometres or 75 miles) made up of volcanic rocks. There is probably a small core of iron, with a radius of approximately 300 kilometres (186 miles).

The surface of the Moon is heavily cratered, probably due to meteorite impact, and the largest (viewed from Earth), is 300 kilometres (186 miles) across and surrounded by gigantic cliffs, (up to 4250 metres or 14,000 feet high). The dark side of the Moon was a mystery until 1959 when *Luna 3*, a Russian space rocket took photographs of it as it flew round. The surface is dry and rocky and in 1969, the American astronauts Neil Armstrong and Edwin Aldrin made the first human footprints on its dusty surface. Other spacecraft have since landed and brought back samples of lunar rock and soil, and analysis of these has revealed that the Moon must be at least 4000 million years old.

A distinctive feature of the Moon is its *maria* (sing. *mare*) which were once thought to be seas. Their origin is not yet established although it is thought that they date from 3300 million years ago.

Interest has recently been rekindled in the Moon with the discovery of ice. This makes it feasible to colonize the Moon and use it as a base for further exploration into space.

moraine a general term for ridges of rock debris deposited by GLACIERS, and marking present or former ice margins, although it originally referred to ridges of debris around alpine glaciers. There are numerous forms of moraine from LATERAL MORAINE (deposited at the side) to terminal moraine (deposited at the leading edge of an active glacier). A medial moraine is a merging of lateral moraines due to the convergence of two glaciers and ground moraine includes a wavy surface of TILL, BOULDER CLAY or glacial DRIFT. There are essentially two types of moraine when considering formation. Dump moraine is where material is literally dumped (during glacier retreat) at the margin or end of a glacier when the ice is stationary. Push moraines denote the edges of ice advance when the ice moves over sediment and pushes up ridges. Large moraines of this type can be seen, often formed annually, and stratified sediments may be faulted and folded due to the force of the glacier.

mordant a chemical that is used in dyeing when the dye will not fix directly onto the fabric. The mordant (mainly weak basic HYDROXIDES of aluminium, chromium and iron) impregnates the fabric, and the dye then reacts with the mordant, thus fixing it to the fabric.

morning star a planet (usually VENUS or possibly MERCURY) seen to the east in the sky around sunrise.

morphine a crystalline ALKALOID occurring in opium. It is used widely as a pain reliever but misuse can be dangerous.

mother liquor the solution which remains after a substance has crystallized out of that solution.

motility the ability to move independently.

mucous membrane tissue which is found as a layer lining cavities in the body which connect with the exterior, e.g. gut, respiratory tracts etc. It is made up of an epithelium (a protective tissue, of closely packed cells, which has additional functions) which contains mucus-secreting cells and often CILIA.

multiple star a star system which contains three or more stars revolving around a common centre of gravity and held in position by their gravitational forces.

mutation a change, whether natural or artificial, spontaneous or induced, in the constitution of the DNA in CHROMOSOMES. Mutation is one way in which genetic variation occurs (see NATURAL SELECTION) since any change in the GAMETES may produce an inherited change in the characteristics of later generations of the organism. Mutations can be initiated by ionizing radiations and certain chemicals.

myelin sheath a fatty substance (myelin) that surrounds axons in the central nervous system of vertebrates and functions as an insulating layer.

mylonite a rock type produced in zones of faulting and shearing.

myoglobin an oxygen-carrying globular protein comprising a single POLYPEPTIDE chain and a haem group. The haem is an iron-containing molecule consisting essentially of a PORPHYRIN which holds an iron atom as a chelate (see CHELATION). The iron can reversibly bind (that is, it can also give it up) oxygen in haemoglobin and myoglobin. The oxygen in myoglobin is released only at low oxygen concentrations, for example, during hard exercise when the muscles demand more oxygen than the blood can supply. Thus myoglobin provides a secondary/ emergency store of oxygen. Mammals such as whales have large amounts of myoglobin in their muscles to facilitate diving.

myosin a large protein found originally with ACTIN in muscles but occurring in most EUCARYOTIC cells. There are at least two varieties, one which is involved in cell locomotion and the other in muscle contraction. The myosin in muscles provides, with actin, the molecular basis of muscle contraction.

N

nacreous clouds a cloud formation at great height before sunrise or after sunset when its colouring is similar to mother of pearl.

nadir the lowest point—the pole which is vertically below the observer in the CELESTIAL SPHERE. It is opposite to the ZENITH.

nanometre a unit of measurement that is used for extremely small objects. One nanometre equals one thousand millionth of a metre (i.e., 10^{-9}.m).

naphtha a mixture of HYDROCARBONS obtained from several sources including COAL TAR and PETROLEUM. Naphtha from coal tar contains mainly AROMATIC hydrocarbons whereas petroleum naphtha hydrocarbons are predominantly ALIPHATIC.

naphthalene consists of two BENZENE rings joined together (with the loss of some carbon and hydrogen—condensed) to give the formula $C_{10}H_8$. It is a white crystalline solid and is obtained from COAL TAR (in the fraction, or part, boiling between 180° and 200°C) and from PETROLEUM fractions, by demethylation i.e. the removal of the METHYL group from methylnaphthalene. Naphthalene is more reactive than benzene and it is used mostly in the production of plasticizers and resins. Some derivatives of naphthalene are important in the manufacture of dyestuffs.

nappe a large-scale geological structure (tens of kilometres) occurring as a sheet of rock which has been pushed over a fault plane or THRUST at the base. They are formed due to compression, but an alternative mechanism is the sliding due to gravity along a low angle fault, as seen in the Swiss Alps. In many cases, the origin of nappes may be due to a combination of both mechanisms.

natural background (see also BACKGROUND) the radiation due to natural RADIO-ACTIVITY and COSMIC RAYS which must be accounted for in the detection and measurement of radiation.

natural gas hydrocarbons in a gaseous state which when found are often associated with liquid petroleum. The gas is a mixture of methane and ethane, with propane and small quantities of butane, nitrogen, carbon dioxide and sulphur compounds.

　　As with petroleum, gas owes its origin to the deposition of sediments that contain a lot of organic matter. After deposition and burial and through the action of heat, with time, oil and gas are produced. These migrate to a suitable reservoir rock where it resides until it is extracted by drilling.

natural glass if MAGMA is cooled quickly, there is insufficient time for crystals to grow from the melt, and a glass is produced. Glasses produced by meteoric impact are found on the Moon. Glasses are called METASTABLE solids, because with time they alter.

natural logarithm see **logarithm**.

natural number any of the set of positive integers also known as the counting numbers: 1, 2, 3, 4 ....

natural selection the process by which evolutionary changes occur in organisms over a long period of time. Darwin explained natural selection (see DARWIN) by arguing that organisms that are well adapted to their environment will survive to produce many offspring, whereas organisms that are not so well adapted to their environment will not. As the better adapted organisms are successfully transmitting their genes from one generation to the next, these are the organisms that are "selected" to survive.

nautical mile the standard international unit of distance used in navigation. One nautical mile is defined as 1852 metres.

neap tides a TIDE of a range up to 30% less than the mean tidal range, which occurs near the Moon's first and third quarters (every 14 days) when the Moon, Earth and Sun are at right angles and the Sun's tidal influence works against the Moon's.

nebula clouds of interstellar gas or dust which was used to describe a collection (GALAXY) of stars.

Neptune is normally the eighth planet of the SOLAR SYSTEM with its orbit between that of URANUS and PLUTO. However, for about twenty years in every 248 years, Pluto's orbit approaches closer to the Sun than Neptune's. At the time of writing, the two planets are within this period and Neptune will be the outermost one until 1999. Neptune is a vast planet, one of the *gas giants*, and is around 4493 kilometres from the Sun. It is extremely cold with surface temperatures of approximately -200°C (-328°F), and the atmosphere consists mainly of methane, hydrogen and helium. The diameter at the poles is 48,700 kilometres and at the equator it is 48,400 km which is about four times that of the Earth. The mass of the planet is 17 times greater than that of Earth and it takes 165 years to circle once around the Sun. Neptune is such a long way from the Earth that it can only be viewed using the most powerful telescopes, and even then it appears to be minute. Two astronomers, John Couch Adams in 1845 in Britain, and Urbain Leverrier in France in 1846, worked out the existence and position of Neptune before it could actually be seen. They noticed that the path of the orbit of nearby URANUS was not as expected, and worked out that it was being affected by the gravitational pull of another planet. A year later, Neptune was viewed for the first time. Neptune spins once on its axis every 18-20 hours and probably has a core of frozen rock and ice.

Most information about Neptune has been obtained from the *Voyager 2* space probe in August, 1989. It took 9000 photographs which show that Neptune is surrounded by a faint series of rings. There is a large dark cloud, the size of the Earth, called the *Dark Spot* which has a spinning oval shape. Winds whip through the atmosphere at velocities of 2000 kilometres per hour and there are white methane clouds which constantly change shape. Neptune has 8 SATELLITES or moons, two of which are large and are called *Triton* and *Nereid*. Triton is the largest moon and is about the same size as Mercury; it orbit is in the opposite direction to that of the other moons. It has a frozen surface with icy volcanic mountains and appears to have lakes of frozen gas and methane. The thin atmosphere of Neptune is mainly nitrogen gas.

neritic zone the zone of shallow water near the seashore which extends from low tide to a depth of approximately 200 metres. Most BENTHIC organisms live in this zone because sunlight can penetrate to such depths. Sediments deposited here comprise sands and clays with features such as RIPPLE MARKS (as seen on beaches).

Nernst, Walther Hermann (1864–1941) a German physical chemist who was awarded the 1920 Nobel prize for chemistry for his proposal of the heat theorem, which was formulated as the third law of THERMODYNAMICS. In 1889, he also developed the Nernst equation, which is used to determine the electromotive force of a cell that contains non-standard constituents.

neuron another name for the nerve cell that is necessary for the transmission of information in the form of impulses along its body. The impulses are carried along long, thin structures of the neuron, known as the axon, and are received by shorter, more numerous structures called dendrites.

neutral a term indicating that a solution or substance is neither acidic or alkaline. The most well-known example of a neutral substance is pure water, which should have a pH of seven.

neutralization a reaction that either increases the pH of an acidic solution to neutral seven or decreases the pH of an alkaline solution to seven. Acid-base neutralizations occur in the presence of an INDICATOR, which undergoes a colour change when the reaction is complete.

neutron an uncharged particle that is found in the nucleus of an ATOM. The MASS of the neutron is 1.675×10^{-24}g, which is slightly larger than the mass of the PROTON. It was discovered by James Chadwick in 1932. Different numbers of neutrons in atoms with the same number of protons result in ISOTOPES.

neutron diffraction the scattering of NEUTRONS by the atoms in a crystal. Depending upon the nature and structure of the crystal, information can be obtained about its atomic and magnetic structure.

neutron star a small body with a seemingly impossibly high density. A star that has exhausted its fuel supply collapses under gravitational forces so intense that its ELECTRONS and PROTONS are crushed together and form NEUTRONS. This produces a star 10 million times more dense than a WHITE DWARF—equivalent to a cupful of matter weighing many million tons on earth. Although no neutron stars have definitely been identified, it is thought that PULSARS may belong to this group.

névé (or firn) a compacted snow, forming as intermediate between snow and glacial ice, and which has survived the melting of a summer.

newton the standard international unit of force. One newton (N) is the force that gives one kilogram an acceleration of one metre per second, per second.

Newtonian telescope a type of telescope in which the image is reflected from a main mirror into an eyepiece on the side of the tube.

Newton's laws of motion the fundamental laws of mechanics, which describe the effects of force on objects. Developed by Isaac Newton (1643–1727), the famous scientist, the three laws of motion state:

First law—any object will remain in a state of rest or constant linear motion provided no unbalanced force acts upon it.

Second law—the rate of change of momentum is proportional to the applied force and occurs in the linear direction in which the force acts.

Third law—every action has a reaction, which has a force equal in magnitude but opposite in its direction.

Until very recently, the above laws could not be directly demonstrated as they are idealized relationships that take place in the less idealized systems here on earth, where the effects of, say, FRICTION have to be considered.

niche all the environmental factors that affect an organism and its community. Such factors include available space, dietary and physical conditions necessary for the survival and reproduction of a SPECIES. Only one species can occupy a specific niche within a community, as the coexistence of any species is subject to distinctions in their ecological niches.

nimbostratus a grey layer of cloud which obscures the Sun and which produces continuously falling rain (or snow).

nitration the addition of the nitro group (NO_2) to organic compounds. It is usually undertaken by reaction with nitric and sulphuric acids. It is used in the manufacture of explosives.

nitric acid a colourless, corrosive acid liquid with the chemical formula HNO_3. It is a powerful oxidizing agent, attacking most metals and producing nitrogen dioxide (NO_2). It is prepared on a large scale by passing a mixture of AMMONIA (NH_3) and air over heated platinum, which acts as a CATALYST. It is used widely in the chemical industry.

nitrides compounds of nitrogen with other elements. Nitrides are prepared by the action of nitrogen or gaseous ammonia. Electropositive elements (Group 1 and 2 metals) form ionic compounds while elements from Groups 3 to 5 (see APPENDIX 1) form a lattice type of structure with nitrogen in the holes. This results in compounds (called refractory hard metals) which are hard, unreactive and have high melting points.

nitrification the process by which bacteria in the soil change ammonia (NH_3) present in decaying matter into nitrite (NO_2^-) and nitrate (NO_3^-) ions. The nitrate ions produced in this way are taken up by plants and are used to make their proteins. Nitrification requires free oxygen, as the bacteria *Nitrosomonas* and *Nitrobacter* oxidize ammonia and nitrites respectively.

nitrogen a colourless, odourless gas which exists as a molecule containing two atoms (N_2) and forms almost 80% of the atmosphere by volume. It is obtained commercially by fractionation (see FRACTIONAL DISTILLATION) of liquid air and is itself used extensively in producing AMMONIA, as a refrigerant, and as an inert, unreactive atmosphere. It is vital for living organisms because it occurs in PROTEINS (and NUCLEIC ACIDS). The NITROGEN CYCLE is the circulation of the element through plant and animal matter into and from the atmosphere.

nitrogenase the enzyme which catalyses (see CATALYST) the production of nitrogenous compounds from free NITROGEN in the process of NITROGEN FIXATION.

nitrogen cycle the regular circulation of nitrogen due to the activity of organisms. Nitrogen is found in all living organisms and forms about 80 per cent of the atmosphere (this proportion is maintained by the nitrogen cycle). The start of the nitrogen cycle can be taken as the uptake of free nitrogen in the atmosphere by bacteria (NITROGEN FIXATION) and the uptake of nitrate (NO_3^-) ions by plants. The nitrogen is incorporated into plant tissue, which in turn is eaten by animals. The nitrogen is returned to the soil by the decomposition of dead plants and animals. NITRIFICATION then converts the decomposing matter into nitrate ions suitable for uptake by plants.

nitrogen fixation the process by which free (available) nitrogen (N_2) is extracted from the atmosphere by certain bacteria. Some free-living bacteria can use the nitrogen to form their AMINO ACIDS, while other nitrogen-fixing bacteria live in the root nodules of leguminous plants (peas and beans) and provide the plants with nitrogenous products. This enables the plant to survive in nitrogen-poor conditions while the bacteria has access to a carbohydrate supply in the plant (this give and take arrangement is called a *symbiotic* relationship). The nitrogen-fixing bacteria are able to convert free nitrogen into nitrogenous products because of the presence of the enzyme nitrogenase within their cells.

noble gases the elements comprising group 8 of the PERIODIC TABLE. Noble gases are usually referred to as INERT GASES because of their relative unreactivity. They are, from lightest to heaviest: helium (He), neon (Ne), argon (Ar), krypton (Kr), xenon (Xe) and radon (Rn). Radon is radioactive and is a potential problem for houses built on areas of granite from which radon escapes.

noble metals metals such as platinum, silver and gold which are highly resistant to attack by acids and corrosive compounds. They tend not to react chemically with non-metals.

noctilucent clouds thin, very high clouds of ice or dust which are sometimes brilliantly coloured because they reflect light from the Sun when it is below the horizon.

node the site on a plant stem from which the bud and leaves grow. In astronomy, when the orbit of an object intersects a plane such as the equator, the two points of intersection are called nodes.

nodule a small swelling/structure on a plant, especially the root which is due to nitrogen-fixing bacteria.

nonagon a nine-sided polygon that, if regular, has equal sides all with interior angles of $140°$.

non-ionic detergents detergents which do not ionize in water to produce large anions or cations.

normal a line perpendicular to the tangent of a curve or contact point of a line or plane.

normal fault a fault where the direction of dip is towards the side which has been moved down.

normal salt any SALT formed by an ACID, which loses more than one hydrogen ion (H^+) per molecule during a NEUTRALIZATION reaction. It should be noted, however, that the production of a normal salt is dependent on the quantities of the acid and BASE used in the reaction.

Northern Lights see **aurora**.

nova (plural *novae*) in the literal sense this is a new star, (nova being Latin for *new*), but it may also be a star that suddenly burns brighter by a factor of five to ten thousand. It seems that a nova is one partner in a BINARY STAR. The smaller star burns much hotter than the sun while the other partner is a vast expanse of hot red mist called a RED GIANT. An explosion results if cooler gas from the red cloud reaches the hot star, causing it to burn up more brightly—a nova. A red giant may eventually become a WHITE DWARF.

nuclear chemistry the study of reactions involving the changes from one type of atom to another due to a nuclear reaction, achieved either by decay or collision with other particles.

nuclear fission is the splitting process which results when a neutron strikes a nucleus of, for example, uranium-235. The nucleus splits into two and releases more neutrons and a lot of energy. A *chain reaction* develops when the neutrons go on to split further nuclei and the energy released becomes enormous. The splitting of the uranium creates two other nuclei, both radioactive:

Uranium-235 + neutron → Krypton-90 + Barium-144 + neutrons + energy

This chain reaction occurs when there is a certain mass of uranium-235, known as the *critical mass*. This was the principle employed in the first *atom bomb*, when two pieces of uranium-235 were brought together to exceed the critical mass and produce a destructive force of unimaginable proportions. Nuclear fission can be controlled by using synthetic rods to absorb excess neutrons. This is the basis of NUCLEAR POWER, which generates the energy used to propel nuclear submarines and to produce electricity from nuclear power plants.

nuclear fusion is where two nuclei are combined to form a single nucleus with an accompanying release of energy. Ordinarily nuclei would repel each other due to the like electrical charge and so very high collision speeds have to be used, which in practice means the use of incredibly high temperatures. A fusion reaction may be:

$$\text{Deuterium} + \text{Tritium} \rightarrow \text{Helium} + \text{neutron} + \text{energy}$$
$$_1^2\text{H} + _1^3\text{H} \rightarrow _2^4\text{He} + _0^1\text{n}$$

and it is necessary to raise the temperature of the hydrogen gas to around 100 million K. Because thermal energy has to be supplied before the nuclear reactions occur, fusion is often called *thermonuclear fusion*. Fusion occurs in the Sun, and in an uncontrolled way in the *hydrogen bomb*, but it is technically very difficult to control in the way that nuclear fission is managed. Research into this subject involves the use of a *tokamak* which employs a magnetic field shaped like a doughnut to trap the hot gas. Thermonuclear reactors could be a solution to energy supply problems because the fuel is readily available and the waste product is not radioactive and it is also inert (unreactive).

nuclear magnetic resonance spectroscopy an analytical technique used for the identification and analysis of organic materials. It has also been developed to form a medical imaging tool. It relies upon the absorption of radio waves by the nuclei of certain atoms.

nuclear power the production of energy from the controlled NUCLEAR FISSION that involves uranium and plutonium as fuels. Nuclear power is used to generate electricity by removing the huge amount of energy released during fuel fission away from the core reactor to the outside, where it is converted to steam and generates electricity by driving turbines. In a *nuclear reactor* heat from the nuclear reactions heats water into steam which drives the turbines. The core of a reactor contains the nuclear fuel which may be uranium dioxide with uranium-235. Neutrons produced by the fission reactions are slowed down by a graphite core to ensure the *chain reaction* continues. The graphite core is called the *moderator*. *Control rods* of boron-steel are lowered into or taken out of the reactor to control the rate of fission. Boron absorbs neutrons and so if rods are lowered there are fewer neutrons available for the nuclear fission and the reactor core temperature will fall. This is a *thermal reactor*. In a *fast breeder reactor* low grade uranium surrounds the core and impact from neutrons creates some uranium-239 which forms plutonium, which itself can be used as a reactor fuel.

nucleic acid a linear MOLECULE that acts as the genetic information store of all cells and which ensures living organisms continue from generation to generation. Nucleic acids occur in two forms, deoxyribonucleic acid (DNA) and ribonucleic acid (RNA), but both forms are composed of four different NUCLEOTIDES, which react to form the long chain-like molecule. DNA is found inside the nucleus of all EUCARYOTES as it is the major part of CHROMOSOMES, but RNA is found outside the nucleus and is essential for protein synthesis.

nucleolus an object enclosed by a membrane found within the NUCLEUS. The nucleolus contains DNA, protein and ribosomal RNA. It is involved in the manufacture of ribosomes and therefore protein synthesis.

nucleophile a reactive molecule that will readily "donate" its unshared pair of electrons. Nucleophiles will attack the low electron density regions of other molecules.

nucleosynthesis nuclear fission reactions occurring in stars and supernovae explosions that produce elements other than hydrogen and helium.

nucleotide a MOLECULE that acts as the basic building block of the NUCLEIC ACIDS, DNA and RNA. The structure of a nucleotide can be divided into three parts— a five-carbon sugar molecule; a phosphate group; and an organic base. The organic base can be either a PURINE, e.g. adenine, or a PYRIMIDINE, e.g cytosine.

nucleus (*plural* **nuclei**) in biology, the ORGANELLE that contains the chromosomes of EUCARYOTIC cells. Molecules enter and leave the nucleus via the pores in the NUCLEAR MEMBRANE. Such molecules include AMINO ACIDS and MESSENGER RNA. The nucleus is usually the largest part of the cell.

   In chemistry and physics, the term nucleus refers to the small, positively charged core of an ATOM that contains the PROTONS and NEUTRONS. The electrons of an atom orbit the nucleus.

   In astronomy, the term has several uses including the central core of a COMET which is made up of ice and dust, and the central part of the galaxy.

numerator the number or quantity to be divided by the denominator of a fraction. For example, the fraction $^3/_4$ has a numerator of 3.

numerical forecasting weather forecasting which relies upon numerous observations and measurements both 'on the ground' and in the atmosphere and the subsequent calculation of what should follow (based also upon natural laws).

nunatak in an area undergoing glaciation, nunataks are rock mountain peaks that poke through the ice.

oblique angle any angle that does not equal 90° (right angle) or any multiple of 90°.

obsidian a volcanic rock which is formed by the rapid cooling of a molten rock with a composition of granite. It is generally glassy, black with a vitreous (glassy) lustre and in composition contains a lot of silica (SiO_2)

occlusion is when warm and cold air meet in a DEPRESSION, where the warm air is lifted above the colder air.

oceanic crust the upper part of the oceanic lithosphere down to the MOHOROVICIC DISCONTINUITY. It is formed of several layers commencing with a top layer of sediment which may be thin or absent (as over oceanic ridges). However, near to continental shelves, (see CONTINENTAL SHELF), sediment may accumulate in thicknesses up to 2 or 3 kilometres. Beneath this is a layer of LAVAS (basaltic) and dykes which together are about 2 km. thick. The next layer is approximately 5 km (3 miles) thick and its composition is similar to that of gabbro, a coarse-grained basic igneous rock. The base of this layer approaches the MANTLE in composition. These layers seem to remain remarkably constant between ocean basins (see *also* MID-OCEANIC RIDGE).

oceanography the study of the oceans, including tides, currents, the water and the sea floors.

oceans technically, those bodies of water that occupy the ocean basins, which begin at the edge of the continental shelf. Marginal seas such as the Mediterranean, Caribbean and Baltic are not classed as oceans. A more general definition is all the water on the Earth's surface, excluding lakes and inland seas. The oceans are the North and South Atlantic; North and South Pacific; Indian and Arctic. Together with all the seas the salt water covers almost 71% of the Earth's surface.

From the shore the land dips away gently in most cases—the continental shelf—after which the gradient increases on the continental slope leading to the deep sea platform (at about 4 km depth). There are many areas of shallow seas on the continental shelf (*epicontinental seas*) e.g. North Sea, Baltic and Hudson Bay. In the ice age, much of the shelf would have been land and conversely should much ice melt, the continents would be submerged further. The floors of the oceans display both mountains, in the form of the mid-oceanic ridges, and deep trenches. The ridges rise 2-3 km from the floor and extend for thousands of kilometres while the trenches reach over 11 km below sea level, at their deepest (Mariana Trench, south-east of Japan).

The oceans contain *currents*, i.e. faster-moving large-scale flows (the slower movements are called *drifts*). Several factors contribute to the formation of currents, including the rotation of the Earth, prevailing winds, differences in temperature and sea water densities. Major currents move clockwise in the northern hemisphere and anticlockwise in the southern hemisphere. Well-known currents include the Gulf Stream and the Humboldt current.

obtuse angle any angle that lies between but does not equal 90° and 180°.

octahedron a geometrical solid that consists of eight triangular faces, and if all are equilateral triangles then the octahedron is said to be regular.

oestrogens a group of female sex hormones, including some sterols, e.g. oestradiol, which is one component of oral contraceptives. Oestrogens are steroid hormones produced mainly by the ovaries and they control sexual characteristics.

ohm (Ω) the unit of electrical resistance. Between two points of an electrical CONDUCTOR, one VOLT (V) is needed to force a current (I) of one AMPERE through a RESISTANCE (R) of one ohm, i.e. $V = IR$. This is known as **Ohm's law** (after the German physicist who formulated it, Georg Simon Ohm [1787-1854]), which can be rewritten: $R = V/I$.

oils are greasy liquid substances obtained from animal or vegetable matter or mineral sources, and they are complex organic compounds. There are also many synthetic oils. There are basically three groups of oils: the *fatty oils* from animal and vegetable sources; *mineral oils* from petroleum and coal; and *essential oils* derived from certain plants. Typical vegetable oils are extracted from soya beans, olives, nuts and maize. The essential oils are volatile (evaporate quickly) and are used in *aromatherapy* and in making perfumes and flavourings. Examples are peppermint oil, clove oil, oil of wintergreen and rose oil.

Mineral oils are actually FOSSIL FUELS and come under the general term of petroleum.

oil shale a dark, fine-grained shale (a type of SEDIMENTARY ROCK) containing organic substances that produce liquid HYDROCARBONS on heating, but do not contain free PETROLEUM.

olivine a rock-forming mineral with the formula $(Mg,Fe)_2SiO_4$. The mineral occurs in IGNEOUS ROCKS that do not contain much silica (e.g. basalts and gabbro) and peridotite which is a coarse-grained ULTRABASIC igneous rock with olivine, PYROXENE and GARNET. It is also found in STONY METEORITES and lunar basalts.

omnivore any organism that eats both plant and animal tissue.

oncogene any gene directly involved in cancer. Oncogenes may be part of a specific VIRUS that has managed to penetrate and replicate within the host's cell, or they may be part of the individual's genetic information which is stored in the chromosomes and which has been transformed by radiation or a chemical.

ontogeny the complete development of an individual to maturity.

oocyte a cell that undergoes MEIOSIS to form the female reproductive cell (egg or ovum) of an organism. In humans, a newborn female already has primary oocytes, which will undergo further development when puberty is reached but which will only complete secondary meiosis to form the secondary HAPLOID oocyte if fertilization occurs.

oolite a LIMESTONE made up largely of ooliths (or ooids). Ooliths are growths of calcium carbonate that look like tiny balls or pellets. They have a small fragment at their core, perhaps a fragment of shell, and this is surrounded by concentric layers of calcium carbonate. Their formation is favoured by shallow marine conditions with raised temperature and salinity.

ooze a deep sea mud made up of clays and the calcareous or siliceous remains of certain organisms e.g. diatoms.

open chain an organic compound with an open chain not a ring structure, as in aliphatic compounds, e.g. ALKANES, ALKENES and ALKYNES and compounds formed from them.

open clusters clusters of loosely gathered stars that have a similar motion through space. There are likely to be a few hundred stars, with gas and dust clouds e.g. the Ursa Major cluster.

optical fibre a small, thin strand of pure glass that uses internal reflection to transmit light signals. Optical fibres are more efficient than conventional cables as they are much smaller, thus requiring less space, and can carry much more data. As such they are replacing metal telephone cables. They are also used in medicine to look inside the body.

optical isomerism the existence of two chemical compounds that are ISOMERS, each being the mirror image of the other.

orbit the path of one heavenly body moving around another which results from the gravitational force attracting them together. The lighter body moves around the heavier one which is itself also in motion. The speed at which a heavenly body travels depends upon the size of its orbit. This is determined by the distance between the two heavenly bodies.

orbitals are a means of explaining the structure of the atom, bonding and similar phenomena. The BOHR THEORY suggested that ELECTRONS were positioned in definite orbits about a central NUCLEUS. However, it was soon discovered that this was too simple and that electrons behave in some ways as waves, which makes their position in space much more imprecise. Hence the old "particle in an orbit" picture was replaced by an electron "smeared out" into a charge cloud or orbital. An atomic orbital is thus one associated with an atomic nucleus and has a shape determined by QUANTUM NUMBERS. Various types of orbital, designated s, p, d, etc, are distinguished. An s orbital is spherical, and a p orbital is dumbbell-shaped. When two atoms form a COVALENT BOND, a molecular orbital with two electrons is formed, associated with both nuclei (see s and π BOND). The overlapping of atomic orbitals in a carbon-carbon single bond (e.g. ETHANE) creates a molecular orbital centred on the line joining the two nuclei. In a carbon-carbon double bond (e.g. ETHENE), the second of the two bonds is created by two overlapping p orbitals, forming the π bond. The two overlapping dumbbells create two sausage-like spaces of electron "cloud" on each side of the line joining the nuclei. BENZENE has torus-shaped (doughnut-shaped) molecular orbitals on each side of the ring due to overlap and merging of p atomic orbitals.

order of magnitude the approximate size of an object or quantity usually expressed in powers of 10.

ordinal number in general, 1st, 2nd etc. rather than 1, 2 etc. (cardinal numbers).

ordinal scale a statistical scale that arranges the data in order of rank in the absence of a numerical scale with regular intervals. Ordinal scales are ideal for data that contain relationships such as bad, good, better, best, as the data can be put in rank order but no regular interval can be measured between the ranked judgements.

ordinate the vertical or y-axis in a geometrical diagram for CARTESIAN CO-ORDINATES. For example, a point with co-ordinates (2,-6) has an ordinate of -6.

ore any naturally occurring substance that contains metals or other compounds that may be used commercially. However, the extraction of the desired metal will only proceed if the process is possible economically and chemically. Some relatively unreactive metals such as copper and gold exist as native ores (that is, as the metal itself) with no need of extraction, but most metals are obtained by extracting them from their oxygen-containing (oxide) ores.

organ a distinct and recognizable site or unit within the body of an organism which consists of two or more types of tissue. It is specialized in terms of its structure and function and, in animals, examples include the kidneys, liver, skin and eyes.

organelle any structure that is bound by a membrane to separate it from the other cell constituents and which has a particular function within the cell. Organelles are found in the cells of all EUCARYOTES and include the CHLOROPLASTS and vacuoles of plant cells in addition to the NUCLEUS, MITOCHONDRIA, GOLGI APPARATUS, and other small vesicles and structures of both animal and plant cells.

organic chemistry the branch of chemistry that is concerned with the study of carbon compounds. Organic chemistry includes studies of the typical bond arrangements and properties of carbon compounds containing hydrogen and, less frequently, oxygen and nitrogen. As most organic compounds are derived from living organisms, two major areas for study are the biologically important organic compounds and the commercially important organic compounds, e.g. ALKANES derived from oil.

organism any living creature including micro-organisms, plants and animals. There are very many different kinds of organism with new species being discovered all the time. At the other end of the scale, over the course of the Earth's history, numerous types of organism have become extinct. *Biotic* is an adjective relating to life or living things (hence *biota*, the plant and animal life of a region). Thus for any organism, the other organisms around it make up the biotic environment.

origin in a graph, the point of intersection of the horizontal (x-axis) and the vertical (y-axis) axes. USING CARTESIAN CO-ORDINATES the origin has co-ordinates of (0, 0).

orogeny a period of mountain building, many of which have occurred in the past, each lasting for millions of years.

oscillation the regular fluctuation of an object whether by means of a cycle, vibration or rotation. In the case of a simple pendulum, oscillation refers to its regular swinging motion and, when used in connection with electrical circuits, oscillation refers to the production of an ALTERNATING CURRENT.

oscillator a CIRCUIT or device that produces an ALTERNATING CURRENT or voltage of a specific FREQUENCY as its output signal. It turns direct current into alternating current. They are used in televisions and radios.

osmosis the process in which solvent molecules (usually water) move through a semi-permeable membrane to the more concentrated solution (i.e. the solution that contains more SOLUTE molecules). This is because there is a difference in size between the molecules of solvent and solute such that only the solvent molecules can pass through the membrane. Many mechanisms have evolved to prevent the death of animal cells either by too much water entering a cell by osmosis, causing it to rupture, or by too much leaving by osmosis, causing it to shrink (called plasmolysis). Such mechanisms include the presence of a pump within the membrane of animal cells, which actively regulates the concentration of vital cellular IONS and the excretion of salt through the gills of marine bony fish to remove the salt gained by diffusion and drinking. Osmosis is an important process in water and mineral uptake by the roots of plants.

outcrop exposure at the Earth's surface of part of a rock formation.

outwash fan a fan-shaped deposit of sands and gravel laid down by glacial melt-water at the margin of boulder clay derived from the glacier.

overburden within a sequence of sediment, the strata which lie over and therefore compress those beneath. Also, superficial material which overlies solid rock.

oviparous in animal reproduction the term describing the development of the embryo and subsequent hatching occurring outside the female's body. Oviparous reproduction is found in birds, most fish, and reptiles.

ovoviviparous the term describing the development of offspring within the body of the female but where there is no development of a placenta. Ovoviviparous reproduction is found in certain species of fish, reptiles and insects, where the young are retained within the mother's body for protection. They continue to receive their nutrients from the egg and not from the mother.

oxbow lake is formed when a river MEANDERS and the formation of meanders becomes such that the river's course follows large, snake-like loops. Eventually the 'neck' between a looped meander is cut and the river straightens its course, leaving a cut-off loop or oxbow lake, shaped something like a horseshoe.

oxidation any chemical reaction that is characterized by the gain of OXYGEN or the loss of ELECTRONS from the reactant. Oxidation can occur in the absence of oxygen, as a molecule is also said to be oxidized if it loses a hydrogen atom.

oxide a compound formed by the combination of OXYGEN with other elements, with the exception of the INERT GASES. Oxides can be divided into acidic oxides (e.g., sulphur dioxide SO_2) which form salts when reacting with a base; basic oxides (such as calcium oxide, CaO) which form salts in a reaction with an acid;

amphoteric oxides which can react both ways (e.g., aluminium oxide, Al_2O_3); and neutral oxides such as CO, carbon monoxide which do not react with acids or gases.

oxidizing agent any substance that will gain ELECTRONS during a chemical reaction. Oxidizing agents will readily cause the OXIDATION of other atoms, molecules or compounds, depending on the strength of the oxidizing agent and the reactivity of the other substance. The following are all examples of oxidizing agents arranged in order of increasing oxidizing strength—sodium ions (Na^+), sulphate ions (SO_4^{2-}), and oxygen molecules (O_2).

oxygen a colourless and odourless gas, which occurs as the molecule made up of two atoms, O_2. It is essential for the respiration of most life forms. It is the most abundant of all the elements, forming 20 per cent by volume of the atmosphere; about 90 per cent by weight of water; and 50 per cent by weight of rocks in the crust. It is manufactured by the FRACTIONAL DISTILLATION of liquid air and is used for welding, anaesthesia and rocket fuels. On heating, it reacts with most elements to form oxides.

ozone a form of oxygen that exists not as the usual O_2 with two atoms of oxygen, but as O_3, with three atoms per molecule. Ozone is more reactive than oxygen and can react with some hydrocarbons in the presence of sunlight to produce toxic substances that irritate the eyes, skin and lungs. Minute quantities of O_3 are found in sea water. It forms the Earth's OZONE LAYER, 15 to 30 kilometres above the Earth's surface.

ozone layer a part of the Earth's atmosphere, at approximately 15-30 km height, that contains ozone. Ozone is present in very small amounts (one to ten parts per million) but it fulfils a very important role by absorbing much of the Sun's ultraviolet radiation, which has harmful effects in excess, causing skin cancer and cataracts and unpredictable consequences to crops, and plankton.

Recent scientific studies have shown a thinning of the ozone layer over the last 20 years, with the appearance of a hole over Antarctica in 1985. This depletion has been caused mainly by the build-up of CFCs (chlorofluorocarbons) from aerosol can propellants, refrigerants and chemicals used in some manufacturing processes. The chlorine in CFCs reacts with ozone to form ordinary oxygen, lessening the effectiveness of the layer. CFCs are now being phased out but the effects of their past use will affect the ozone layer for some time to come.

P

palaeogeography the study of the physical geography at periods in the past.

palaeontology is the scientific study of FOSSILS which allows information to be collected on how organisms lived, what they looked like and how they evolved with time. Fossils are found in sedimentary rocks and occur throughout most of the geological record.

palynology the study of (mainly fossil) pollen, spores and some other microfossils. It deals primarily with structure, classification and distribution and is used in petroleum exploration, and palaeoclimatology (the study of ancient climates). In the main, the spores and pollen are highly resistant and in some sedimentary sequences are the only fossils that can be used for stratigraphic correlation (matching one rock with a similar rock elsewhere which is of the same age).

parabola a plane curve traced out by a point moving so that its distance from a fixed point (focus) is equal to its perpendicular distance from a fixed straight line (directrix). It is also the curve made by cutting a cone with a flat plane that is parallel to the side of the cone that slopes. Mirrors with a parabolic shape are used in searchlights and telescopes because incoming parallel rays of light are reflected onto the focus. This property is used in a telescope, and thus the reverse applies in searchlights.

parabolic velocity in astronomy, the velocity a body would need to make a parabola about the centre of attraction (also called ESCAPE VELOCITY).

paraffins the general term for the ALKANES, saturated ALIPHATIC HYDROCARBONS with the formula C_nH_{2n+2}. They are quite unreactive hence the name paraffin (from the Latin *parum affinis*, little allied).

parallax the apparent movement in the position of a heavenly body due in fact to a change in the position of the observer. It is therefore caused, in reality, by the Earth moving through space on its orbit. The distance of a heavenly body from Earth can be calculated by astronomers using parallax. The direction of the body from Earth is measured at two six month intervals when the Earth is at either side of its orbit. From the apparent change in position, the distance of the body from Earth can be deduced. A similar situation arises when looking at objects in a room. If you move your head sideways, objects some distance away are unaffected, whereas objects near to your eyes appear to move.

parallel circuit an electrical CIRCUIT in which each component has the same PO-TENTIAL DIFFERENCE across it (measured in volts) but a different amount of current flowing through it. The amount of current flowing through each component depends upon a phenomenon called RESISTANCE, and in parallel circuits the total resistance (R) of any circuit components is given by the following relationship:

$$1/R = 1/R1 + 1/R2 + 1/R3 ... etc.$$

If a circuit divides into branches then the different current through the branches together are the same as the current in the main circuit.

parallelogram a four-sided POLYGON, which has opposite sides that are parallel and equal in length. Parallelograms can have four sides all of equal length (equi-lateral parallelogram, i.e. a rhombus), four equal angles (equiangular parallelo-gram, i.e., a rectangle) or have all four sides and angles equal, i.e. a square.

parameter an arbitrary constant or variable that determines the specific form of a mathematical equation, as a and b in $y = (x - a)^2 + b$. Changing the value of the parameter results in various outcomes. More simply, the parameters of a rectangle are its length and height.

parasite is an organism which obtains its food by living in or on the body of another living organism, without giving anything in return. The organism on which the parasite feeds is known as the *host*. Usually, the parasite does not kill the host as its future depends upon their mutual survival. However, sometimes a host species can become so seriously ill or weakened by the presence of the parasite that it dies. Parasites are of two kinds: *ectoparasites* attach themselves to the host's surface or skin and examples include the blood-sucking headlice, ticks and fleas. *Endoparasites* live inside the host's body, often within the gut or muscle, and examples are tapeworms, roundworms and liver flukes. Parasitic plants usually twine themselves around their host. Mistletoe is a partial parasite which obtains water and minerals from the host, although it performs its own PHOTOSYNTHESIS.

parenthesis (*plural* **parentheses**) the curved brackets ( ) used to group terms or as a sign of aggregation in a mathematical or logical expression.

parsec an astronomical unit of distance used for measurements beyond the so-lar system. It equals 206,265 ASTRONOMICAL UNITS or 3.26 light years, which is almost 31 million million kilometres.

parthenogenesis the development of a new individual from an unfertilized egg. Parthenogenesis is most common among the lower invertebrates, such as insects. The process can be part of the honey bee life cycle if the eggs laid by the queen remain unfertilized by sperm. The larvae from these eggs will develop into the male bees (drones), whose only function is to produce sperm. If the eggs laid by the queen are fertilized then the larvae develop into sterile female worker bees or fertile queens, depending on the food supply. Parthenogenesis also occurs in some plants, such as the common dandelion.

partial fractions the simple FRACTIONS into which a larger fraction may be separated so that the sum of the simpler fractions equals the original fraction.

particle accelerator is a machine for increasing the speed (and therefore the kinetic energy) of charged particles such as protons, electrons and helium nuclei by accelerating them in an electric field.

Accelerators are used in the study of subatomic particles. To split an atom, particles travelling close to the speed of light are required.

The first accelerator was built by Cockcroft and Walton in 1932, and with it they split the atom for the first time. The energies of modern machines are measured in GeV (gigaelectron-volts, i.e. billions) and at the Fermi Laboratory in the USA, 800 GeV has been reached. Higher energies can be achieved using a *storage ring* which is a toroidal (doughnut-shaped) component in some accelerators. Particles enter the ring and can stay there for many weeks or months. A ring 300 metres in diameter in Geneva produces energies up to 1700 GeV.

particle size (grain size) the diameter of the grains in a SEDIMENTARY ROCK. The size is determined by physical means—sieving for the smaller grains and direct measurement for the larger. It is customary to size small grains by referring to the diameter of a sphere with the same volume.

pascal the unit of PRESSURE named after the French mathematician, philosopher and physicist, Blaise Pascal (1623–1662). One pascal (Pa) is defined as the FORCE of one NEWTON acting on a square metre, i.e. $1 \text{ Pa} = 1 \text{ Nm}^{-2}$. One atmosphere of pressure (760mm mercury, the air pressure at sea level) is approximately 100 kilopascals (kPa).

Pascal's law of fluid pressures this law states that the PRESSURE of a fluid is the same at every point since any pressure that may be applied will be transmitted equally to all points of the vessel containing the liquid. Pascal discovered this principle while mountaineering with his father. He realized that the column of mercury in the barometers he carried would vary in length—essentially the principle behind all hydraulic systems.

Pascal's triangle an array of numbers in the shape of a pyramid starting with one, such that each number is the sum of the two numbers in the row directly above it. The result is a triangle of potentially infinite size, the beginning of which is as follows:

$$
\begin{array}{c}
1 \\
1 \quad 1 \\
1 \quad 2 \quad 1 \\
1 \quad 3 \quad 3 \quad 1 \\
1 \quad 4 \quad 6 \quad 4 \quad 1 \\
1 \quad 5 \quad 10 \quad 10 \quad 5 \quad 1 \\
1 \quad 6 \quad 15 \quad 20 \quad 15 \quad 6 \quad 1
\end{array}
$$

Pascal's triangle is an extremely useful method for determining the COEFFICIENTS when using the BINOMIAL THEOREM. Interestingly, the total of each horizontal row is a power of 2, the fifth row being 16 which is 2^4.

passive immunity is when an individual can resist disease using ANTIBODIES that have been donated by another individual rather than by producing its own antibodies. Passive immunity is obtained by young mammals from their mother's milk during the first few weeks of life as the newly born are virtually incapable of antibody production. In humans, breast-fed infants will receive most of their maternal antibodies from their mother's milk, but one antibody, IgG, will be found in all infants, whether breast-fed or bottle-fed, as IgG can move across the placenta during development of the foetus.

Pasteur, Louis (1822–1895) a French chemist and bacteriologist who was the first to demonstrate that a colony of bacteria could be grown on an appropriate culture medium that had been infected with a few cells of the micro-organism. This experiment demonstrated that living cells had an inheritance of their own and helped to discredit the theory of SPONTANEOUS GENERATION of life. Although Pasteur had been aware of the role of micro-organisms in FERMENTATION since 1858, he did not accept their role in causing disease until several years later. He demonstrated that certain forms of micro-organisms could be used in inoculation, providing immunization for the host, and in 1885 he produced the first rabies vaccine.

pasteurization is the process named after the French chemist and biologist, Louis Pasteur. It involves the partial sterilization of food and kills potentially harmful bacteria. Milk is an obvious example. If it is heated to 62°C for 30 minutes, it kills bacteria that could cause tuberculosis and it increases the shelf life by delaying fermentation because other bacteria have also been killed or damaged. An alternative treatment involves heating milk to 72°C for 15 seconds. Higher temperatures still are used to produce 'long-life' milk. Pasteurization is also used with beer and wine, to eliminate any yeast which would create cloudiness in the drink.

pathogen is any organism that causes disease in another organism. Most pathogens that affect humans and other animals are bacteria or viruses, but in plants there is also a wide range of fungi that act as pathogens.

Pauling, Linus Carl (1901–) an American biochemist who worked out the structure of crystals of simple molecules and pure proteins by using X-ray crystallography. With fellow colleagues, Pauling discovered one of the regular structures common to all proteins, the a-helix and, using ELECTROPHORESIS, isolated the abnormal HAEMOGLOBIN that causes an hereditary form of anaemia.

peat an organic deposit formed from compacted dead and possibly altered, vegetation. It is formed from vegetation in swampy hollows and occurs when decomposition is slow due to the oxygen-free (ANAEROBIC) conditions in the waterlogged hollow. Sphagnum moss is among the principal source-plants for peat. As peat builds up each year, water is squeezed out of the lower layers causing the peat to shrink and consolidate. Even so, cut peat has a high moisture content and is dried in air before burning. In Ireland and Sweden, peat is used in power stations.

pectins complex polysaccharides (see SACCHARIDES) that occur in the cell walls of certain plants, particularly fruits. They are soluble in water and acid solutions, and gel with sucrose, that is they set to a jelly, and so they are useful in jam making.

pedology the study of soils—their composition, occurrence and formation.

peduncle in botany, the main stalk of a plant bearing several flowers. In zoology, the stalk by which certain organisms such as brachiopods, anchor themselves to the substrate.

pelagic a descriptive term for organisms living in the sea between the surface and middle depths. Pelagic sediments (e.g. OOZE) are deep water deposits made up of minute organisms and small quantities of fine-grained debris.

pelitic rock a METAMORPHIC ROCK formed by the metamorphism of shales and mudstones. The particular aluminium-silicate minerals formed will depend on the conditions of metamorphism but usually includes MICA.

pendulum a device that consists of a weight swinging on the end of a wire of very little mass, suspended from a fixed point. The time of a complete swing is given as $T = 2\pi\sqrt{(l/g)}$ where l is the length of the wire and g the acceleration due to gravity which indicates that it depends not on the size of the weight, but the length of the wire. A pendulum is in equilibrium, that is, stable when the weight is directly underneath the point of fixing.

penecontemporaneous a term used to describe any process happening in a rock soon after its formation.

pentad a period of five days which is used in meteorological records because it is a fraction of a normal year.

pentagon any plane shape that has five sides. A regular pentagon has sides of equal length and five interior angles each measuring $108°$.

pentahedron any three-dimensional figure that has five plane faces.

pepsin an enzyme which is secreted by cells lining the stomach and which breaks down protein. It is active in acid conditions and by catalysing the process of hydrolysis, helps the breakdown of proteins to give PEPTIDES and AMINO ACIDS.

peptide bond a covalent linkage formed when two AMINO ACIDS join together. As all amino acids have a common molecular structure, the reaction always involves the elimination of a water molecule as the amino group (NH_2) of one amino acid molecule joins to the carboxyl group (COOH) of another molecule. A number of amino acids joined together in this way forms a POLYPEPTIDE.

periastron when a body is orbiting a star, the periastron is the point at which the body is nearest to the star.

perigee see **apogee**.

perihelion similar to PERIASTRON but referring to the Sun, that is, the point when a body's orbit takes it closest to the Sun. This applies to planets, comets, spacecraft, etc.

perimeter the total distance around the outside of a closed plane figure, such as the circumference of a circle.

period the time taken for a body to complete one full OSCILLATION, which can involve a vibration, rotation or harmonic motion. Period (T) has seconds (s) as its units, and it is the reciprocal of FREQUENCY, i.e. $T = 1/f$.

In chemistry, periods are the horizontal rows in the PERIODIC TABLE, for example, those elements between an alkali metal and the next inert gas. The periods are thus hydrogen (H) and helium (He); two periods of few elements, lithium (Li) to neon (Ne), and sodium (Na) to argon (Ar); the two long periods containing the TRANSITION ELEMENTS, running from potassium (K) to krypton (Kr) and rubidium (Rb) to xenon (Xe); the period caesium (Cs) to radon (Rn) and the unfinished period beginning with francium (Fr). A geological period is the second order of geological time e.g., the Carboniferous Period (see APPENDIX 5).

periodic table is an ordered table of all the ELEMENTS arranged by their ATOMIC NUMBERS, i.e. the number of protons and electrons in an atom. The arrangement means that elements with similar properties are grouped near to each other. The horizontal rows are called *periods* and the vertical rows are *groups*. Elements with the same number of electrons in their outer shell behave in a similar way and this is the basis of the vertical group. Moving from left to right along the periods corresponds to the gradual filling of successive electron shells and an increase in the size of the atom.

There are various sections within the periodic table as follows:

alkali metals	Li, Na, K, Rb, Cs, Fr
alkaline earth metals	Ca, Sr, Ba, Ra
chalcogens	O, S, Se, Te, Po
halogens	F, Cl, Br, I, At
nert gases	He, Ne, Ar, Kr, Xe, Rn
rare earth elements	Sc, Y, La to Lu
lanthanides	Ce to Lu inclusive
actinium series	Ac onwards
transuranium elements	elements after U
platinum metals	Ru, Os, Rh, Ir, Pd, Pt

Mendeléef was a Russian chemist who constructed the first periodic table, but based upon atomic weights. This basic principle, modified to use atomic numbers rather than atomic weights formed the basis of the modern table and even then, in 1869, allowed Mendeléef to predict the existence of undiscovered elements.

peristalsis the involuntary muscular contractions responsible for moving the contents of tubular organs in one direction. Peristalsis occurs in the alimentary canal of animals as the alternate waves of contraction and relaxation of smooth muscle move food and waste products along.

permafrost frozen ground which is permanently frozen except for surface melting in the summer. It is defined as when the temperature is below 0° for two or more consecutive years. About one quarter of the Earth's land surface is affected, and although it may be very thick (several hundred metres) the larger depths are probably relics from the last ice age. It occurs north of the Arctic Circle in Canada, Alaska and Siberia. Permafrost can cause considerable engineering problems particularly when the heat generated by towns etc. creates some thawing, leading to subsidence and slumping of previously solid ground.

permeability in geology, it is the ability of a sediment, soil or rock to allow the flow of fluids (here taken as gas, oil or water). Specifically and technically, the permeability (also called the hydraulic conductivity) is the volume flow rate of water through a section of porous medium under the effect of an hydraulic gradient (i.e., a difference in water content from one area to another) at a certain temperature.

In physics, it is the diffusion rate of a liquid or gas through a porous material, under the effects of a pressure gradient.

permutation a particular arrangement of a set of objects given by the formula $n!/(n - r)!$, where ! stands for FACTORIAL, n is the number of objects taken r at a time. For example, the number of permutations of four coloured beads from a choice of ten is:

$$\frac{10}{(10 - 4)!} = \frac{10!}{6!} = 5040$$

perpendicular any line or plane that meets another line or plane at a right angle (90°). If the perpendicular is formed by a line meeting a plane or the tangent to a curve, then the line is referred to as the NORMAL of that plane or curve.

perturbation slight changes in the movement of planets from their orbits because of gravitational attraction, drag etc.

petrifaction the process by which organic remains are replaced by, usually, minerals, but in which the original structure is retained (see also FOSSILIZATION).

petrochemicals are chemicals made from crude oil (petroleum) and natural gas which are used to manufacture an enormous range of compounds and materials including plastics, drugs, fertilizers, solvents and detergents. Over 90% of synthetic organic materials come from these sources.

Petroleum fractions

methane and ethane	natural gas
propane and butanes	liquefied petroleum gases
light naphtha	
naphtha	motor spirit (gasoline)
kerosine	jet fuel
gas oil	diesel fuel
heavy distillates	feedstocks for lubricants, waxes etc.
bitumen/asphalt	

petrography the description of rocks, the minerals present and the textures, through study of hand specimens and very thin sections of the rock which is viewed under the microscope.

petroleum (crude oil) a liquid mixture of hydrocarbons that occurs naturally and is formed by the decay of organic matter under pressure and high temperatures. The oil formed migrates from its source (origin) to a permeable reservoir rock where it remains if it is capped or sealed by an impermeable cover. The composition of the petroleum varies with the source. It has to be processed (refined) and is separated initially by FRACTIONAL DISTILLATION into its major components or fractions (gas, liquids, wax, and residues such as bitumen). The liquids include petrol, paraffin oil, and other hydrocarbon liquids. CRACKING is used to break down some substances to create smaller molecules that can be used more readily. In addition to the production of various fuels, petroleum is the basis of the vast PETROCHEMICALS industry. See *also* FOSSIL FUELS.

petrology the study of rocks and in particular their origin, occurrence, mineral and chemical composition and any processes of alteration which they may have undergone. The term is usually applied separately to each class of rock type i.e. sedimentary, igneous or metamorphic petrology.

pH is the measure of a solution's acidity or alkalinity and is the concentration of hydrogen IONS (H^+) in an aqueous solution. The pH is the negative LOGARITHM (to base 10) of H^+ ion concentration, calculated using the following formula:

$$pH = \log_{10}(1/(H^+))$$

The scale of pH ranges from 1.0 (highly acidic, such as concentrated hydrochloric acid), with decreasing acidity until pH 7.0 (NEUTRAL) and then increasing alkalinity to 14 (highly alkaline such as sodium hydroxide). As the pH measurement is logarithmic, one unit of pH change is equivalent to a tenfold change in the concentration of H^+ ions. The pH of solutions is checked using indicators.

phagocytosis a process used by simple one-celled organisms to take in food particles. The particle binds to the cell's surface and is then completely engulfed by a bud formed by the plasma membrane of the cell. This process is also used by certain LEUCOCYTES to engulf and destroy bacteria and old, broken cells.

phase the term used for the change in shape of the bright surface of the Moon due to the relative positions of the Sun, Moon and Earth. It may also apply to the planets and is due to bodies shining only by reflected light from the Sun.

A chemical phase is a part of a system which is chemically uniform but occurs in a different form, for example there are three phases in a system that contains ice, water and water vapour. However, all gases show one phase since they all mix with each other.

phenol has the formula C_6H_5OH and was known as carbolic acid. As a solution in water, it is corrosive and poisonous. It is used as a disinfectant (with the typical "carbolic" smell) and in the manufacture of dyes, explosives, pharmaceuticals and PLASTICS.

phenocrysts if a rock sample shows large, well-formed crystals amongst a mass of smaller crystals, the larger ones are called phenocrysts. An IGNEOUS ROCK showing two sizes of crystals in this way is said to be porphyritic.

phenotype the detectable characteristics of an organism, that is, its appearance which is determined by the interaction between its GENOTYPE (that is, its gene makeup) and the environment in which the organism develops. Organisms with identical genotypes may have different phenotypes, due to development in environments that differ in, for example, the availability of important nutrients or specific stimuli. It is unlikely, however, that organisms that have identical detectable phenotypes will have different genotypes unless they are HETEROZYGOTES. The expression of the dominant gene masks the presence of a recessive gene, as only the expressed gene affects the organism's phenotype.

pheromone a molecule that functions as a chemical signal enabling communication between individuals of the same species. Pheromones are used extensively throughout the animal kingdom and have a wide range of functions. They can act as sexual attractants (very common in insects) and can help establish territories, as demonstrated by the frequent urination by dogs. They function in a way that is similar to hormones and they are small molecules, active in tiny quantities.

phloem tissue in plants that has the major task of transporting food materials from the points of production, the leaves, to areas where they are needed e.g., growing points. The phloem is made up of sieve tubes which are hollow and lie parallel to the length of the plant. The tubes are formed from sieve cells, end to end, with the walls at the ends of cells broken down to permit movement of the metabolites.

phospholipids biological compounds that resemble fats. They contain two FATTY ACIDS joined to glycerol, to which is also joined a phosphate group (PO_4). The phosphate also bonds to a nitrogenous base (hydrocarbon with nitrogen). The fatty acids form the water-hating (hydrophobic) tails and the rest is the water-loving (hydrophilic) head. Phospholipids occur in cell membranes.

phosphorescence LUMINESCENCE that continues after the initial cause of excitation. The substance usually emits light of a particular WAVELENGTH after absorbing ELECTROMAGNETIC radiation of a shorter wavelength.

phosphorus a non-metallic element (symbol P) occurring in several forms (red, white and black), the latter being a high temperature and pressure variety. Phosphorus occurs naturally as compounds, mainly as calcium phosphate $(Ca_3(PO_4)_2)$. It is manufactured by heating the phosphate with sand and carbon in an electric furnace. It is commonly found in minerals and living matter and is vital to life being the main constituent of animal bones. It is used industrially in the manufacture of fertilizers, matches and in organic synthesis.

photic zone the uppermost layer of a lake or sea where there is adequate light to allow PHOTOSYNTHESIS to proceed. The limit will vary on the quality of the water and the material held in suspension, but can be as much as 200 metres.

photobiology the study of the effect of light on living organisms.

photochemistry the study of the effects of radiation (mainly the visible and ultraviolet parts of the SPECTRUM) on chemical reactions. Only light which is absorbed can have any effect and the first stage in a photochemical reaction is the absorption of (a quantum of) light energy by an atom which is then raised to an excited state.

Infrared radiation is ineffective but radiation from the far ultraviolet is strong enough to break chemical bonds. The light absorbed may catalyse a reaction (see CATALYST) or render possible a reaction that would otherwise not proceed. A reactant usually absorbs the light, but where the energy is passed on to a reactant by another species, the process is called photosensitization.

photochromics the term used for materials which are sensitive to light. In some cases materials darken in bright light and the change is reversed when the light source is removed. Certain materials are used in optical memory devices.

photolysis the decomposition or reaction of a substance due to the absorption of light or ultraviolet radiation. Flash photolysis is a technique for studying very fast reactions involving ATOMS or RADICALS in the gas phase. The reactants are subjected to an intense, but brief, flash of light which causes dissociation (the breakdown into ions). Subsequent flashes are used immediately afterwards to identify intermediates which are produced in the reaction (by studying the absorption spectra).

photosphere the visible surface of a star which, in the case of the Sun, is several hundred kilometres thick. It is estimated to have a temperature of 6000K and it is in this zone that sunspots, flares and similar features are seen.

phototaxis the movement or reaction of an organism in response to light (see *also* CHEMOTAXIS).

photon a QUANTUM or packet of energy that is a basic part of all ELECTROMAGNETIC WAVES. Photons are used to explain the quantum theory of light, where the properties of light are explained in terms of particles (photons), as opposed to the wave theory of light, where its properties are explained by the propagation of a wave and how it disturbs a medium. The energy of a photon is proportional to the FREQUENCY of the light beam.

photosynthesis a complex process by which plants make their food, carbohydrates, using water; carbon dioxide (CO_2) and light energy. The end product is glucose and a byproduct is the release of oxygen. Photosynthesis occurs in two stages, known as the CALVIN CYCLE and the LIGHT REACTIONS of photosynthesis. For photosynthesis to occur, an organism must contain light-trapping pigments, which capture light energy in the form of PHOTONS and use the photons to start a series of energy-transfer reactions. Some blue-green algae (cyanobacteria) and the CHLOROPLASTS of all plants contain the essential light-trapping pigment called CHLOROPHYLL that makes them capable of photosynthesis. Photosynthesis is an essential process for regulating the atmosphere as it increases the oxygen concentration while reducing the CO_2 concentration.

phototropism a growth movement shown by parts of plants in response to the effect of light. Plant shoots display positive phototropism as they grow towards the incoming light, but the roots tend to display negative phototropism as they grow away from the light source. Phototropism is caused by the unequal distribution of a plant growth hormone called auxin. This substance occurs in a higher concentration in the darker side of the plant and thus increases growth on this side by producing cell elongation, and the plant then 'bends' towards the light.

phylum (*plural* **phyla**) a part of the classification structure of the animal kingdom. A phylum includes one or more classes that are closely related. Examples are Protozoa (mainly micro-organisms), Arthropoda (e.g., insects and spiders), and Chordata which includes the vertebrates. In the classification of plants, the term *division* is used.

physical chemistry the study of the link between physical properties and chemical composition and the physical changes caused by chemical reactions.

physics is the study of matter and energy, and changes in energy without chemical alteration. Physics includes a number of topics such as magnetism, electricity, heat, light and sound (see *individual entries*). The study of modern physics also includes QUANTUM THEORY, atomic and nuclear physics i.e., subatomic particles and their behaviour (see ELEMENTARY PARTICLES) and the physics of NUCLEAR FISSION AND FUSION. As the research into topics has expanded over recent years, so new subjects begin to develop often on the boundaries of two major disciplines. This has happened in geophysics (geology and physics), biophysics (biology and physics) and astrophysics, which combines astronomy with physics.

phytochemistry the study of the chemical make-up of plants.

pi (π) bond the COVALENT BOND formed when two atoms join to form a molecule. Pi bonds are discussed in terms of molecular ORBITALS, in which the shared electrons orbit the whole molecule rather than an atom. Pi bonds hold the molecule together by forming two regions of electron density above and below an axis between the bonded nuclei of the two atoms.

piedmont glacier an extension of ice from a valley glacier which projects beyond its valley walls onto the adjacent flat plain at the foot of the mountains (that is, the area called the piedmont). Because the ice is now at a lower altitude, it may melt and reduce in size more rapidly than it would normally.

piedmont gravels coarse deposits of pebbles etc. found on the flat lowlands (piedmont). The deposits are laid down by fast-flowing mountain rivers that carry a lot of sediment because they are moving so quickly. Upon reaching the flatter ground, the river slows and cannot carry as great a load and so the larger, heavier material is deposited.

piezoelectric effect this is an effect developed in certain crystals whereby opposite charges are generated on opposite crystal faces by the application of pressure. It occurs because a small charge on the atoms in the crystal moves under pressure. QUARTZ is such a crystal. Uses of this phenomenon include the crystal microphone and quartz watches in which a crystal vibrates rapidly and the electric signal is used to keep time. The opposite also applies in that the application of an electric current alters the shape of the crystal.

pilot balloon a small hydrogen-filled balloon used in meteorology to determine wind speed and direction at high altitude.

pipette originally a glass tube with a fine tip and sometimes a bulbous central portion, used for obtaining a given volume of solution. It is filled by sucking or applying negative pressure by means of a rubber bulb. Automatic pipettes are now more commonplace.

pituitary gland see **endocrine system**.

place value notation when a number has more than one digit, the position, or place value, of each digit in the number is used to indicate what it is worth. For example, using the decimal system, the 6 in 362 and in 3620 stands for something different: in 362 it means 6 "tens" and in 3620 it means 6 "hundreds."

placers (or placer deposits) deposits rich in mineral ores, e.g. platinum, gold, cassiterite (tin oxide). They are produced by the mechanical action of weathering on an original rock, creating a concentration of these minerals because the lighter ones have been removed.

Planck's constant the proportionality constant (h) used in the equation to define the energy of a QUANTUM. Planck's constant has a value of 6.6262×10^{-34}Js and is named after the German mathematician and physicist Max Planck (1858–1947), who proposed the theory that radiant energy consisted of quanta.

planet the name given originally to seven heavenly bodies which were thought to move among the stars which were themselves stationary. The term now applies to those moving in definite orbits about the SUN which, in order of distance, are: MERCURY, VENUS, EARTH, MARS, JUPITER, SATURN, URANUS, NEPTUNE and PLUTO. Mercury and Venus are termed the inferior planets and Mars to Pluto the superior planets, the latter because they revolve outside the Earth's orbit.

planetary nebula a layer of glowing gas produced by and surrounding an evolved star (see HERTZSPRUNG-RUSSELL DIAGRAM) and representing a late stage in star evolution. (It is nothing to do with a planet.)

planetoid see **asteroid**.

plankton very small organisms, often microscopic and including both plants and animals, which drift in the currents of oceans and lakes. The plants (or *phytoplankton*) consist mainly of single-celled algae (such as *Bacillariophyta*) called *diatoms* which photosynthesize (see PHOTOSYNTHESIS) and form the basis of food chains. The animals (or *zooplankton*) include the *larval* stages of larger organisms, some protozoans (Kingdom Protista) and small creatures called copepods which are related to crabs. Plankton provide food for many larger animals, e.g. whales, and are of vital importance in the FOOD CHAIN.

plasma in biology, the same as BLOOD PLASMA. In physics, a plasma is a high temperature gas of charged particles (ELECTRONS and IONS) rather than neutral atoms or molecules. A plasma is electrically neutral overall, but the presence of charged particles means that it can support an electric CURRENT. It is of significance to the study of controlled NUCLEAR FUSION.

plasmid a structure consisting of DNA (with just a few genes) that exists outside the CHROMOSOME in a procaryotic (see PROCARYOTE) cell and is able to replicate independently. Plasmids enable the cell to resist antibiotics, metabolise special compounds as nutrients and perform other special functions.

plasmolysis see **osmosis**.

plastics is a group name for mainly synthetic organic compounds which are mostly POLYMERS, formed by polymerization, that can be moulded when subjected to heat and pressure. There are two types: *thermoplastics* (e.g. PVC or polyvinyl chloride) which become plastic when heated and can be heated repeatedly without changing their properties; and *thermosetting plastics* such as phenol/formaldehyde resins that lose their plasticity after being subjected to heat and/or pressure. Plastics are moulded and shaped while in their softened state and then cured by further heat (thermosetting e.g. epoxy resins, silicones) or cooling (thermoplastics e.g. perspex and polythene).

The first synthetic plastic was *Bakelite*, invented in 1908. Since then the plastics industry has become vast and an enormous range of domestic, leisure, industrial and commercial items are now produced. Plastics can be shaped by blow moulding, vacuum forming, extrusion and injection moulding and they are used extensively in composite materials and laminates.

plate (see also PLATE TECTONICS) in geology the concept of plates arose from the observation that large areas of the crust have suffered little distortion and yet have travelled many kilometres. Thus plates have little seismic (related to earthquakes) or volcanic activity but are fringed by margins which exhibit earthquakes, volcanism and mountain chains (whether submarine or subaerial). There are six major lithosphere plates: Eurasian, American, African, Indo-Australian, Antarctic and Pacific, and numerous smaller plates (e.g. Caribbean and Philippine) and microplates (e.g. the Hellenic in the eastern Mediterranean).

platelet see **blood**.

plate tectonics see **continental drift**.

platinum metals a block of six TRANSITION ELEMENTS with similar properties—ruthenium (Ru), rhodium (Rh), palladium (Pd), osmium (Os), iridium (Ir) and platinum (Pt) (see also PERIODIC TABLE). The platinum metals are commonly found together, with gold and silver. Of the group, platinum is of greatest importance. It is very stable and is used mainly for jewellery, special scientific equipment, and chemical electrodes. It is also used as a CATALYST (see also ZEOLITES).

The Pleiades the open cluster of stars in the Taurus constellation. The seven main stars, each individually named, form this well-known group, although only six are visible when viewed unaided. The Pleiades is about 400 light years away and is roughly 30 light years across.

plug (or **neck**) the cylinder-like remnant of a volcano which is formed by the cooling and solidification of magma within the main feeder of the volcano. Plugs are commonly roughly circular in shape, steep-sided and vary in size from a few metres to over a kilometre. In addition to magma, they may contain PYROCLASTICS. The whole feature is produced by the erosion and removal of the rocks surrounding the plug. It is also known as puy, after Le Puy in the Auvergne region of France.

Pluto the ninth and smallest planet in the SOLAR SYSTEM and the one that lies farthest away from the SUN. The existence of Pluto was predicted by an American astronomer called Percival Lowell from the behaviour of the orbits of its closest neighbours, NEPTUNE and URANUS. Pluto was finally spotted in 1930, fourteen years after Lowell died. Pluto appears as a tiny speck when viewed from Earth and little is known about it, but it probably has an iron core and a rocky surface with a covering of methane ice. Since it is so far from the Sun (a maximum of 7,338 million kilometres), it must be extremely cold, in the region of -230°C. A day on Pluto lasts for almost seven Earth days and a year is 248.4 Earth years. At the equator, the planet has a diameter in the region of 3,500 kilometres. Pluto has a wide elliptical orbit which sometimes brings it closer to the Sun. In 1989, it was at its closest point to the Sun (called its *perihelion*), but this only occurs every 248 years. During this phase of its orbit, Pluto apparently has a thin atmosphere composed of methane gas. However, when it moves away from the Sun again it is possible that this becomes frozen once again. In 1979 it was discovered that Pluto has one small satellite (which was named *Charon*), and is about a quarter the size of Pluto. The two bodies effectively form a double planet system.

podsol a soil with minerals leached (taken out in solution) from its surface layers into lower layers. Podsolization is an advanced stage of leaching which involves the removal of iron and aluminium compounds, humus and clay minerals from the topmost horizons and their redeposition further down.

point of inflection a point where a plane curve changes from a concave to a convex shape, relative to some fixed line, i.e., the point where it "crosses its tangent."

polar axis the diameter of a sphere which passes through both poles. Also, in an equatorial telescope, it is the axis (parallel to the Earth's) the telescope revolves around so that the object is kept in the field of view.

polar co-ordinates the position of a point in space as represented by the co-ordinates (r, q), where q is the angle between the positive x-axis and a line from the origin to the point, and r the length of that line.

polar covalent bond a bond between atoms that is basically covalent with sharing of electrons. However, the sharing of their ELECTRONS is unequal, which leads to the bond, and thus the molecule formed, having partial charges. Polar covalent bonds are formed between atoms that differ in their ability to attract electrons, with one atom having a greater ELECTRONEGATIVITY than another. For example, a hydrogen chloride molecule (HCl) has a polar covalent bond, as the chlorine atom is more electronegative than the hydrogen atom. The result is that the chlorine end of the bond has a denser electron cloud because of its greater attraction for electrons. This causes the chlorine end of the molecule to have a partial negative charge, and as the whole HCl molecule is neutral, the hydrogen end has a corresponding partial positive charge.

Polaris the brightest star in the constellation Ursa Minor, once much used for navigation (in the northern hemisphere).

polarization is when light is made to vibrate in one particular plane. Light normally consists of waves vibrating in many directions and because it is electromagnetic radiation there is an electric and a magnetic field vibrating at right angles to each other. If light is polarized, the electric field vibrations are confined to one plane (*plane polarized light*) called the plane of vibration and the magnetic vibrations are in one plane at right angles, the plane of polarization.

Polarization occurs when light passes through certain crystals (quartz, calcite) or is reflected from some surfaces (e.g. the sea). A polarizer produces polarized light and a polaroid filter is one such material used in sunglasses. Polarized light also has uses in mechanical engineering to reveal stress patterns in materials. Only transverse waves such as light can be polarized; longitudinal waves such as sound cannot.

pollen pollen grains are the male sex cells (*gametes*) of flowering PLANTs. They occur in small pollen sacs contained within a structure called the *anther*, which is part of the male reproductive organ of the flower. The anther occurs on the end of a thin stalk called the stamen.

pollen analysis (see *also* PALYNOLOGY) a useful tool in studying the history of the flora of an area. Because the outer layer of pollen grains is resistant, particularly if deposited under the anaerobic conditions of rapid sedimentation, or in peat or stagnant water, they are widely distributed and preserved. Such analysis contributes to studies of climate change and sediment dating.

polygon a closed plane figure with three or more straight line sides. Common polygons are figures such as the triangle, quadrilateral and pentagon. A square is an example of a regular polygon, one in which all sides and all angles are equal.

A general equation exists for the sum of the interior angles of a polygon with n sides: sum of the interior angles = $180°$ $(n-2)$.

polyhedron a solid figure composed of four or more polygonal plane faces. The more faces it has, the closer it is to a sphere. A cube is an example of a regular polyhedron as all the faces and all the angles of a cube are equal.

polymer a large, usually linear molecule that is formed from many simple molecules called *monomers*. Synthetic polymers include PVC, Teflon, polythene and nylon while naturally-occurring polymers include starch, cellulose (found in the cell walls of plants) and rubber (see *individual entries*). Early versions of polymers were modified natural compounds, e.g. the vulcanization of rubber by heating with sulphur. The first fully synthetic polymer to be developed was *Bakelite* (see PLASTICS) followed by urea-formaldehyde and alkyd resins in the late 1920s. Polyethylene (that is polythene) was first produced by ICI on a commercial basis in 1938 and the Dupont company in America produced the first nylon in 1941. Many synthetic polymers are produced from alkenes (see HYDROCARBONS) in reactions called addition polymerizations which are rapid and require only relatively low temperatures. Condensation polymerization is another means of producing polymers e.g. the nylons (polyamides) and the silicones (polysiloxanes) in which some molecule (often water) is removed at each successive reaction stage.

polynomial in mathematics, an algebraic expression consisting of one or more variables raised to a positive or zero integrated power. For example, a linear polynomial (highest power of x being 1) is $3x + 1$; a cubic polynomial would be $5x^3 + 3x^2 + 6x + 1$ because the highest power of x is 3.

polypeptide a single, linear MOLECULE that is formed from many AMINO ACIDS joined by PEPTIDE BONDS. Polypeptides differ greatly in the number of amino acids they contain (usually from 30 to 1000). Although there are only 20 different amino acids, there are a huge number of possible arrangements in a polypeptide or PROTEIN, as the amino acids can be in any order. Most proteins consist of more than one polypeptide rather than a single polypeptide chain.

population types the two categories of stars within the galaxy. Population I stars which are relatively young, include hot blue stars found in the arms of spiral galaxies. Red stars which occur in globular clusters and the central parts of galaxies make up the older Population II.

porosity essentially the spaces in a rock. However, not all spaces will be connected to each other to permit fluids to pass through the rock.

porphyrins naturally occurring pigments including CHLOROPHYLL and the haem part of HAEMOGLOBIN.

positron a particle with the same mass as the ELECTRON but a positive electrical charge. Positrons are produced during decay processes (see BETA DECAY) and themselves are annihilated on passing through matter (see ANTIMATTER).

potassium (symbol K) is a soft alkali metal which is silver-white and highly reactive. Due to its high reactivity it occurs only combined with other elements. It occurs widely: as potassium chloride (KCl) in sea water; in silicate rocks as alkali FELDSPAR; in blood and milk and also in plants. Potassium is used primarily and extensively in its compounds as fertilizers, potassium hydroxide is used in batteries and ceramics, and alloyed with SODIUM it can be used as a heat transfer medium as a coolant in reactors.

potential difference the work done in driving a unit of electric charge (one COULOMB) from one point to another in a current-carrying CIRCUIT. The unit of potential difference is the VOLT, and the potential difference is frequently referred to as the VOLTAGE.

potential energy a body has potential energy if and when it has been moved to a position from which it can do work when released. A body is said to have potential energy (U) when it has been raised from a resting point A against gravitation to resting point B. The potential energy of such a body can be worked out from $U = mgh$, where m is the mass of the body, g is 9.8 ms^{-2} (GRAVITY), and h is the distance moved. If the body is released from resting point B, its potential energy is transformed into the energy of motion, KINETIC ENERGY. Potential energy is present in a spring that has been stretched and can also be stored in the form of chemical or electrical energy.

power the rate at which work is done by or against a FORCE. Power is also regarded as the rate at which energy is converted from one form to another. The unit of power is the WATT (W), which is equal to the transfer of one joule per second, i.e. $1W = 1Js^{-1}$. Electrical power can be calculated by multiplying the voltage by the current, $P = IV$.

power notation the use of a small number (an EXPONENT) placed next to an ordinary number to show how many times the ordinary number is multiplied by itself e.g.:

$$3^5 \text{ ("3 to the power 5") means } 3 \times 3 \times 3 \times 3 \times 3$$

pozzolana a volcanic soil deposit which, when mixed with lime, produces a cement. It was first used by the Romans when the material was discovered near

Pozzuoli, close to Naples. Pozzolanas are only formed when volcanic activity has been explosive, producing a vitreous PYROCLASTIC material. The term is now used collectively for all materials exhibiting reactivity with lime and which set in the presence of water.

precessional motion is the type of motion shown by a gyroscope. A rotating body precesses when the application of a couple (that is, two equal and opposite parallel forces) with an axis at 90° to the rotation axis causes the body to turn around the third common perpendicular axis.

precipitation in a reaction between two solutions precipitation is the formation of an insoluble substance (precipitate). It occurs because the IONS of the two substances exchange partners. Thus, if a solution of silver nitrate, $AgNO_3$, is mixed with a solution of sodium chloride, NaCl the insoluble silver chloride AgCl forms a precipitate. From the chemical equation for this reaction:

$$AgNO_3(aq) + NaCl\ (aq) \longrightarrow AgCl\ (s) + NaNO_3\ (aq)$$
$$(aq = aqueous,\ s = solid)$$

it can be seen that the chloride ion (Cl^-) has displaced nitrate.

In meteorology, precipitation is the collective term for rain, hail or snow falling from clouds onto the surface of the Earth.

pressure is the FORCE exerted on the unit area of a surface. The pressure of a gas is equal to the force that its molecules exert on the walls of the containing vessel, divided by the surface area of the vessel. The pressure of a gas will vary with its temperature and volume, as stated by BOYLE'S LAW, CHARLES' LAW, and the GAS LAWS. At any depth, the pressure in a liquid or in air equals the weight above the unit area, and therefore as the depth increases, the pressure also increases. This is also the reason for air pressure decreasing as height above sea level increases. The unit of pressure is the PASCAL, although air pressure is commonly measured using mercury BAROMETERS and hence has units of millimetres of mercury (mm Hg) corresponding to the varying mercury levels as air pressure changes.

pressure gradient the rate of change of (atmospheric) pressure on the ground as shown by ISOBARS.

pressure law see **gas laws**.

prime number a number that can only be divided by itself and 1. The first ten prime numbers are 2, 3, 7, 11, 13, 17, 19, 23, 29.

prism in mathematics, a solid with equal and parallel POLYGONS as ends and parallelograms as sides. In physics, a prism is triangular in shape and made of transparent material and light passing through is deviated and split into its component colours. Prisms are used commonly in optical instruments or laboratory experiments.

procaryote (or prokaryote) any organism that lacks a true-membrane NUCLEUS and is either a bacterium or a blue-green algae (cyanobacteria). Procaryotes have a single CHROMOSOME with much less DNA than EUCARYOTES and do not undergo MEIOSIS or MITOSIS. Procaryotes reproduce by a form of asexual reproduction, called binary fission, in which the two sister chromosomes are attached to separate regions on the cell membrane, which starts to fold to form a cleavage. The cell eventually forms two daughter cells after the CYTOPLASM has been completely split by the fusion of the enfolding cell membrane.

progesterone a STEROID hormone, which in mammals is important in pregnancy. It prepares the inner lining of the uterus for the fertilized egg to be implanted, and during pregnancy maintains the uterus and prevents any more eggs being released.

prognostic chart a weather forecast chart specifying the expected meteorological conditions.

prophase the first stage of MEIOSIS or MITOSIS in cells of EUCARYOTES. During prophase, the CHROMOSOMES condense and can thus be studied using a microscope.

protease a group of ENZYMES that act as catalysts (see CATALYSIS) in the breaking up of PROTEINS into PEPTIDES and AMINO ACIDS. Examples are PEPSIN and trypsin.

protein a large group of organic compounds of vital significance to all living matter. They consist of carbon, oxygen, hydrogen and nitrogen and have high MOLECULAR WEIGHTS and comprise hundreds or thousands of AMINO ACIDS joined to form POLYPEPTIDE chains. The amino acid sequence gives to each protein its particular properties. Proteins are used for support, storage, as messengers and carriers of other substances. ENZYMES are another important group of proteins. There are thousands of different proteins in the human body, each with a unique structure but all made up from the same 'pool' of amino acids.

proteolysis the splitting of proteins through the catalytic action (see CATALYST) of PROTEASES. To break down a protein completely (into its amino acids) usually requires several proteases acting one after the other.

proton a particle that carries a positive charge and is found in the NUCLEUS of every ATOM. As an atom is electrically neutral, the number of protons equals the number of negatively charged ELECTRONS. Although the MASS of a proton (1.673 ×

10^{-27} kg) is far greater than the mass of an electron (9.11×10^{-31} kg), their charges are equal in size. The number of protons in the nucleus of an atom (ATOMIC NUMBER) is identical for any one element and is used to classify elements in the PERIODIC TABLE. For example, as every oxygen atom contains 6 protons, it has an atomic number of 6 in the periodic table, whereas every gold atom contains 79 protons and thus has an atomic number and periodic table position of 79.

Proxima Centauri the star nearest to the Sun (about 4.3 light years away) and a member of the constellation Centaurus.

pseudopodium (*plural* **pseudopodia**) the temporary projection from the body of certain cells. Pseudopodia are formed in simple, single-celled organisms, such as *amoeba*, as a mechanism for locomotion and food intake. They are also formed by white blood cells, which use PHAGOCYTOSIS to ingest particles.

PTFE (polytetrafluoroethene) a thermosetting PLASTIC produced by the polymerization of tetrafluoroethene (CF_2CF_2). Under its trade names of Teflon and Fluon, it is used to line saucepans, where its chemical unreactivity and heat resistance are useful. It is also used in engineering applications.

pulmonary artery one of the two arteries that carry deoxygenated blood from the HEART to the lungs, where it is oxygenated. The pulmonary arteries are the only ones that carry blood with a high concentration of carbon dioxide rather than a high concentration of oxygen. All other arteries carry oxygenated blood to the tissues, where oxygen is exchanged for carbon dioxide.

pulmonary vein one of the four veins that carry oxygenated blood from the lungs (two veins leave both the left and right lungs) to the left ATRIUM of the HEART. The pulmonary veins are unique, as they carry oxygenated blood while all other veins carry deoxygenated blood back to the heart after it has exchanged oxygen for carbon dioxide in the various tissues of the body.

pulsar pulsars are thought to be collapsed, rotating NEUTRON STARS which are left after a SUPERNOVA explosion. A pulsar is a source of radio frequency radiation which is given out in regular short bursts. The radiation is in the form of a beam which sweeps through space as the pulsar rotates, and is detected on Earth if Earth happens to lie in its path. The first pulsar was found to emanate from the *Crab Nebula* which is a supernova.

pumice an acidic rock, usually of PYROCLASTIC origin, which occurs as a vesicular frothy glass formed by ejection from a volcano followed by rapid cooling.

purine one of the two different families of structures that form the base components of DNA and RNA. A purine has a double ring structure that consists of both carbon and nitrogen atoms with hydrogen. The bases, ADENINE and GUANINE, are both purines that will form hydrogen bonds with their complementary PYRIMIDINE bases to form the double helix of the DNA molecule.

putrefaction the breakdown (decomposition) of plants and animals after death by anaerobic bacteria.

PVA (polyvinyl acetate) a PLASTIC produced by the polymerization of vinyl acetate. It is used in coatings, adhesives and inks.

PVC (polyvinyl chloride or polychloroethene) the most widely used of the vinyl PLASTICS formed by POLYMERIZATION of vinyl chloride (chloroethene H_2CCHCl). PVC is used for pipes, ducts, mouldings and as a fabric in clothing and furnishings.

pyrheliometer a device for the measurement of direct solar radiation.

pyrimidine one of the two different structures that form the base components of DNA and RNA. A pyrimidine has a single ring, consisting of both carbon and nitrogen atoms (and oxygen, and hydrogen). The bases, cytosine, THYMINE and URACIL, are all pyrimidines.

pyrite otherwise known as fool's gold is iron sulphide, FeS_2. The sulphide mineral occurs widely distributed in numerous environments. It forms a minor mineral in IGNEOUS ROCKS and SEDIMENTARY ROCKS, particularly black shales (which were deposited in ANAEROBIC conditions). It also occurs in replacement deposits and some metamorphic rocks. It was mined for the production of sulphuric acid, but no longer.

pyroclastic rocks rocks, such as pumice, that are formed by the violent expulsion of rock and lava from volcanic vents.

pyroxenes in geology, an important group of rock-forming minerals which are silicates of iron, magnesium and calcium, sometimes with aluminium. Certain varieties contain sodium or lithium. All are characterized by the Si_2O_6 chain structure and there are many varieties due to replacement of one metal by another. The group can be divided into two on their crystal systems but both show very good CLEAVAGES. Pyroxenes are widely distributed in igneous and metamorphic rocks and show a variety of colours, but usually dark greens, brown or black.

pyruvate a colourless liquid formed as a key intermediate in the metabolic process of GLYCOLYSIS and the production of ATP.

Pythagoras' theorem the geometrical theorem that states that in any right-angled triangle, the square of the HYPOTENUSE is equal to the sum of the squares of the two shorter sides. This theorem is named after the Greek philosopher and mathematician of the 4th century BC. For a given right-angled triangle in which the sides are x and y units long, the hypotenuse (h) can be obtained from $h^2 = x^2 + y^2$. Pythagoras' theorem provides a method of calculating the length of any side of a right-angled triangle if the lengths of the other two sides are known.

Q

quadrilateral any geometric shape that has four sides, e.g., a rectangle, square, kite, parallelogram, rhombus.

qualitative analysis the chemical examination of a sample to discover what substances are present.

quantitative analysis the examination of a sample to discover the amounts of the substances present.

quantum (*plural* **quanta**) a small amount or unit of electromagnetic radiation which can be thought of as a particle of energy.

quartz one of the most common rock-forming minerals, SiO_2, which is found in many different kinds of rock. It also forms some semi-precious st   e.g. amethyst and agate.

quasar one of a number of star-like heavenly bodies which are very distant from Earth and which give out light. They are extremely compact and give out light even though they are vast distances away, up to 10^{10} light years.

quotient the result arrived at when a mathematical quantity is divided by another quantity—the 'answer'.

quotient rule a mathematical method used in CALCULUS.

R

radar (**ra**dio **d**etection **an**d **r**anging) the use of radio waves to discover the presence and distance of an object. Radar is used in the navigation of aircraft, ships, missiles and SATELLITES.

radiation the giving out of energy from a source, which may be in the form of ELECTROMAGNETIC WAVES (radio, light, X-rays, infrared rays, etc.), particles or sound waves.

radical a group of atoms within a compound, which are not able to exist on their own and are not changed when the substance is involved in chemical reactions.

radioactivity the giving out of particles (known as α or β particles and γ rays) by unstable substances that are disintegrating.

radio astronomy the recording and study of radio waves given out by many bodies in space including the Sun, stars and QUASARS.

radiocarbon dating a method of dating ORGANIC material up to 8000 years old. There is a small amount of radioactive carbon 14 in the atmosphere which is taken up naturally by plants and animals during life. When the organism dies there is no more uptake of ^{14}C and it starts to decay, with a HALF-LIFE of 5730 years. A date for the sample is arrived at by comparing the amount of ^{14}C left with known standards.

radiochemistry the scientific study and purification of radioactive materials.

radiography the method or process of making an image of an object on photographic film (or on a fluorescent screen), using X-rays (or similar rays such as gamma rays). The photograph produced is called a *radiograph* and the use of X-rays in this way is widely used in hospitals, for example, to discover if bones are fractured or to detect the presence of a tumour.

radiometric dating a precise method of dating rocks that measures the amounts of certain radioactive elements present in their original and decayed states.

radio (waves) a means of communication through space, sending information in the form of sound, pictures and digital data.

radiolysis the chemical breakdown or decomposition of a substance when it is placed in the path of IONIZING radiation.

radionuclide any ISOTOPE of an element that undergoes natural radioactive decay.

radiosonde an instrument that is used in meteorology and that is carried through the different levels of the atmosphere by a balloon. The apparatus measures temperature, HUMIDITY and pressure and results are sent to a radio receiver.

radio telescope an instrument used to detect and analyse electromagnetic radiation coming from various sources in space.

rain the condensation of water vapour into droplets when moist air is cooled below its DEWPOINT. As the air cools it can hold less moisture and some condenses as rain.

rainbow the characteristic display of colours formed by the REFRACTION and REFLECTION of sunlight by raindrops in the air.

rain gauge an apparatus for measuring rainfall which may be simply made up of a funnel leading into a collecting bottle.

rain shadow a dry, or even desert-like area of land that occurs behind a mountain range which runs parallel to the sea. It occurs because most of the moisture carried by winds from the sea, condenses and falls on the mountain slopes which face the ocean, as the air rises and cools. By the time the air reaches the far side of the mountains, it is nearly dry, hence the lack of rain and desert-like condition of the land. An example in the USA is the wet western side of the coastal range and Sierra Nevada mountains, compared with the much drier parts of eastern California and Nevada.

raised beach an ancient beach that is now above the level of the shoreline, often because there has been a fall in sea level.

ratio numbers these are used to compare the sizes of two or more quantities. If, in a class of 24 pupils, there are 8 boys and 16 girls, the ratio of boys to girls is $8 : 16$ or $1 : 2$. It is usual to try and reduce one of the numbers to one.

rational number a number that can be arrived at by dividing one quantity by another quantity, including all whole numbers and most fractions.

rayon the term applied formerly to "artificial silk," but now to two man-made cellulose fibres, viscose and cellulose acetate rayon.

reagent a chemical substance or solution that is used to produce a characteristic reaction in chemical analysis.

real number any RATIONAL or IRRATIONAL number. Real numbers exclude imaginary numbers and COMPLEX NUMBERS.

recessive allele a gene form that is not expressed (that is, it does not show as a feature in the organism) and will therefore not affect the PHENOTYPE of the organism unless the organism is HOMOZYGOUS for that particular recessive allele.

Although an organism that is HETEROZYGOUS for a recessive allele will possess this allele, the dominant form of the gene will be expressed, thus masking the presence of the recessive form.

reciprocal the inverse ("other way up") of a FRACTION; the reciprocal of a number A is $1/A$. For example, the reciprocal of $5/12$ is $12/5$, and the reciprocal of 6 is $1/6$.

recombinant DNA a new DNA sequence formed by the insertion of a foreign DNA fragment into another DNA molecule. Recombinant DNA is used extensively throughout GENETIC ENGINEERING, when bacteria are frequently used as hosts for the expression of recombinant DNA molecules and the subsequent coding for the desired protein. It is particularly useful for producing a significant quantity of a human PROTEIN, such as INSULIN.

recombination see **genetic recombination**.

rectangle see **parallelogram**.

rectifier a device that converts ALTERNATING CURRENT into direct current. It is basically a diode (which allows current through in only one direction) which allows forward flow but not backward flow of current.

recurring decimal when a number or set of numbers is, after a certain point, repeated indefinitely. The recurring figures are signified by dots often, for example, 0.3 is equivalent to 0.333333 ...Also called repeating decimal.

red blood cell see **erythrocyte**.

red giant an ageing star that is extremely hot and has used up about 10% of its hydrogen. The outer layers are cooler than the intensely hot centre. As the name implies, these are very large stars. The red giant, *Aldebaran*, is 35 times larger than the SUN with a diameter of 50,100,000 kilometres. Some red giants go on to become WHITE DWARFS, using up their hydrogen at an increased rate. The sun will eventually become a red giant in about 5000 million years time.

red shift the light observed from certain galaxies shows a displacement of spectral lines towards the red end of the spectrum - hence red shift. This is interpreted as being due to the DOPPLER EFFECT and signifies that the galaxies are receding into space.

redox potential a method for evaluating the REDUCTION or OXIDATION potential of a reactant. Redox potentials are arranged on an arbitrary scale, which uses the standard hydrogen electrode as the reference redox reaction by assigning it a potential of zero volts. The strongest REDUCING AGENTS, i.e. those most easily oxidized, are at the top of the scale while the strongest OXIDIZING AGENTS, i.e. those most easily reduced, are at the bottom of the scale.

redox reaction a chemical reaction in which both REDUCTION and OXIDATION are involved. If the overall REDOX POTENTIAL for such a reaction has a positive value, then it is a spontaneous and feasible reaction.

reducing agent any substance that will lose ELECTRONS during a chemical reaction. Reducing agents will readily cause the REDUCTION of other atoms, molecules or compounds, depending on the strength of the reducing agent and the reactivity of the other reactant. The strongest reducing agents are active alkali metals such as lithium (Li), potassium (K), barium (Ba), and calcium (Ca).

reduction any chemical reaction that is characterized by the loss of oxygen or by the gain of ELECTRONS from one of the reactants. A molecule is also said to be reduced if it has gained a hydrogen atom. There is always simultaneous OXIDATION if reduction has occurred in any reaction.

reflecting telescope a telescope which uses a mirror to focus light rays. There are several versions including the NEWTONIAN TELESCOPE, named after Isaac Newton who first realized its potential. The largest telescopes in the world are all of the reflecting variety.

reflection the property of certain surfaces whereby rays of light falling upon them are returned (reflected) in accordance with definite laws. The incoming or incident ray becomes the reflected ray.

reflux is when a liquid is boiled in a flask and a condenser (a tube with a cooled inner portion upon which the vapour condenses) is attached so that vapour condenses and is returned to the flask. This keeps the liquid boiling but prevents loss by evaporation.

refracting telescope a telescope which uses lenses to focus light rays, first applied to astronomy by Galileo.

refraction is the bending of, most commonly, a light ray when it travels from one medium to another, e.g., air to water. The refraction occurs at the point where the light passes from one material to another and is caused by the light travelling at different velocities in the different media. The incident (ingoing) ray passing into a material becomes the refracted ray and in an optically more dense medium it is bent towards the normal to the interface (the line at right to the join between the different materials). The two angles of incidence and refraction are related by *Snell's Law* which states that the ratio of the sines (see TRIGONOMETRY) of the two angles is constant for light passing from one given medium to another. The value of the ratio of the sines of the angles is called the *refractive index*, measured when light is refracted from a vacuum into the medium.

refractive index (n) the ratio of the SINE of the angle of incidence (the angle between the incident ray and the line drawn PERPENDICULAR to the surface at that point) to the sine of the angle of refraction, when light is refracted from a vacuum into the medium.

regeneration the repair or regrowth of the bodily parts of an organism that have been damaged and have been subsequently lost. Regeneration is rare in higher, complex animals but it is quite common in the lower, simpler animals in which the extent of regeneration can range from limb regeneration in crustaceans to the regeneration of the whole organism from one segment, as in certain annelid worms. Regeneration is common in plants and occurs naturally, as in VEGETATIVE PROPAGATION, or can be induced to propagate plants of economic importance, such as the potato and tobacco plants. Complete regeneration of any plant is only possible if its vegetative cells have retained the full genetic potential enabling them to replicate every part of the plant.

regional metamorphism metamorphism involving both pressure and temperature and possible shear stress due to converging plates. The extent of orogenic belts (see OROGENY) means the effects are on a *regional* scale. The metamorphism can occur as several events tied in with several tectonic events and can occur before, during or after a tectonic episode. The rock FABRIC changes with metamorphic grade (increasing in grain size) and certain groups of minerals are generated relating to pressure/temperature régimes.

regolith a fine powdery covering on the MOON (and other planets and asteroids), created by meteoritic impact. On the Moon, the material is several metres thick. In geology it refers to the layer of unconsolidated and weathered material which lies over solid rock. It may comprise rock fragments, mineral grains and soil components and in the humid tropics can reach enormous thicknesses—commonly tens of metres.

regression in mathematics, the connection between the expected value of a random VARIABLE and the values of one or more possibly related variables. In biology, the tendency to an average state from an extreme one.

relative atomic mass (*formerly called* **atomic weight**) the mass of atoms of an element given in atomic mass units (u), where $1u = 1.660 \times 10^{-27}$ kg.

relativity the theory derived by EINSTEIN that establishes the concept of a four-dimensional space-time continuum where there is no clear line between three-dimensional space and independent time, hence space and time are considered to be bound together. The important results of the theory include the understanding that the mass of a body is a function of its speed; the derivation of the mass-energy equation, $E = mc^2$ (where c = speed of light), and the relative nature of time itself, i.e. there is no absolute value or interval of time.

replication the duplication of genetic material, generally before cell division.

reproduction the production of new individuals of the same species either by *asexual* or *sexual* means. The term usually refers to sexual reproduction which involves the joining together (called *fusion*) of special sex cells, one of which is female (the egg or *ovum*) and the other is male (e.g. *sperm* in animals and pollen in flowering plants). The sex cells of any organism are known as *gametes*. Many organisms produce gametes within special reproductive organs. In the flower of a seed plant, the male sex organs are the stamens which produce pollen. The *carpels* are female and produce *ovules* which later develop into seeds after *fertilization*.

reservoir rock a porous and permeable rock which can hold oil, gas or water. Typical rock types are SANDSTONE, LIMESTONE or DOLOMITE. Almost two thirds of oil occurrences are in sandstones and related rock types with almost one third in carbonate rocks. For example, dune sands in the Permian which now lie under the southern North Sea, the Netherlands and north Germany form the reservoir for gas from the Coal Measures beneath.

residual deposits the production of rock waste (from clays to boulders) due to weathering, or the weathered material remaining after soluble components have been dissolved out. In both cases the processes occur *in situ*.

resins natural resins are organic compounds secreted by plants and animals e.g. rosin, derived from pine trees. Synthetic resin is the term now applied to any synthetic PLASTIC material produced by polymerization.

resistance (R) measured in OHMS and calculated as the potential difference between the ends of a CONDUCTOR, divided by the CURRENT flowing. Apart from superconductors, materials resist the flow of current to varying degrees, and some of the electrical energy is converted to heat.

resistivity the reciprocal of a material's conductivity, giving the resistance in terms of its dimensions.

resistor a component of electric CIRCUITS, used to provide a known RESISTANCE.

respiration the process by which living cells of an organism release energy by breaking down complex organic compounds (food molecules) into simpler ones using enzymes. Respiration can occur in the presence of oxygen (AEROBIC RESPIRATION) as in most organisms, or in its absence (ANAEROBIC RESPIRATION) and has an initial stage called GLYCOLYSIS, which is common to both forms of respiration. The term respiration is also used, although less frequently, for gaseous exchange (better known as breathing) in an organism, which involves the uptake of oxygen from, and the release of carbon dioxide to, its surrounding environment.

retrovirus *see* **virus.**

reversible reaction a chemical reaction which can proceed in either direction. The incomplete reaction results in a mixture of reactants and products and the balance can be altered by a change in the controlling factors, whether pressure, temperature or concentration.

rheostat a RESISTOR of variable RESISTANCE.

rhesus see **blood grouping**.

rhombus see **parallelogram**.

ribosomal RNA (rRNA) one of the three major classes of RNA, which is transcribed from DNA in a structure of eucaryotic nuclei called the NUCLEOLUS. Along with many PROTEINS, ribosomal RNA forms the cellular structures called RIBOSOMES, which are found in both EUCARYOTIC and PROCARYOTIC cells.

ribosome the structure within the cell that is the site of PROTEIN synthesis in all EUCARYOTIC and PROCARYOTIC cells. Ribosomes are composed of one large and one small sub-unit, which contain RIBOSOMAL RNA and associated proteins. Analysis of procaryotic and eucaryotic ribosomes indicates that they share the same evolutionary origins as their structure, and the RNA they contain (except a segment unique to eucaryotes) are virtually identical. Ribosomes assemble at one end of a MESSENGER RNA molecule and move along the molecule to build the POLYPEPTIDE chains of all proteins in a process called TRANSLATION.

Richter scale the scale used to measure the intensity of earthquakes which uses the AMPLITUDE of seismic waves, which depends on the depth of the earthquake focus. Recording stations register the waves and for a shallow earthquake the magnitude is given by:

$$M = \log (a/t) + 1.66 \log\Delta + 3.3$$

where a is the maximum amplitude, t the period (the time between a repeat of the same wave form) and Δ is the angular distance between the focus and the station. A slightly modified version is used for deeper earthquakes. Earlier systems of intensity measured, e.g. that devised by Mercalli, the Italian seismologist relied more upon the effects seen or felt by observers when the seismic waves reached them. Below in brief is the arbitrary scale from 1 to 12:

1	*Instrumental*	detected only by seismographs
2	*Feeble*	noticed by sensitive people
3	*Slight*	similar to a passing lorry
4	*Moderate*	rocking of loose objects
5	*Rather strong*	felt generally

6	*Strong*	trees sway; loose objects fall
7	*Very strong*	walls crack
8	*Destructive*	chimneys fall; masonry cracked
9	*Ruinous*	collapse of houses where ground starts to crack
10	*Disastrous*	buildings destroyed; ground badly cracked
11	*Very disastrous*	bridges and most buildings destroyed; landslides
12	*Catastrophic*	total destruction; ground moves in waves

right angle an angle of 90°.

right ascension a coordinate used, with others, in specifying positions on the CELESTIAL SPHERE.

ring compound see **closed-chain**.

ripple marks the preservation of ripples in sandstones which may exhibit small cross-laminations reflecting current movement.

RNA (*abbreviation for* **ribonucleic acid**) a NUCLEIC ACID that exists in all living cells. It is involved in the synthesis of protein. Genes from DNA control the production of a type of RNA (messenger RNA, mRNA) which, with the special structures, controls synthesis of a particular polypeptide.

rock an aggregate of minerals or organic matter. Rocks are classified into thee types depending upon the way in which they were formed: IGNEOUS, SEDIMENTARY and METAMORPHIC.

ROM (read only memory) the part of the memory in a computer which is fixed and can be read but not written to or altered.

rubber a natural hydrocarbon POLYMER (polyisoprene) from the *Hevea brasiliensis* tree. Items made from rubber are produced by adding various agents followed by vulcanization (heating in the presence of sulphur). Synthetic rubbers are polymers (or copolymers) of simple molecules.

ruminant any mammal that has four compartments in the stomach to aid the digestion of large amounts of plant material. Ruminants include cattle, sheep, deer and giraffes. In the first section of their stomach, the rumen, food is enveloped in a mucus and is partially digested by an ENZYME called cellulase, which is supplied by the billions of bacteria living in the rumen. After this, the food is brought back to the mouth (regurgitated) and, after chewing, it passes through a further two sections (where water is removed) and it eventually ends up in the true stomach (the abomasum) which contains the enzymes needed for complete digestion.

S

sabkha a flat, coastal belt situated between desert dunes and a lagoon or the sea. It is a site for the formation of evaporite deposits, notably CARBONATES and sulphates. It is named after the Trucial Coast in the Persian Gulf, *sabkha* being the Arabic word for salt flat.

saccharides SUGARS (and therefore CARBOHYDRATES) divided into mono-, di-, tri- and polysaccharides. Monosaccharides are the basic units, simple sugars; disaccharides, e.g., sucrose and lactose, are formed by condensing two monosaccharides and removing water. Sucrose gives, on HYDROLYSIS, a mixture of glucose and fructose; trisaccharides comprise three basic units and polysaccharides are a large class of natural carbohydrates including STARCH and CELLULOSE.

saccharin a white, crystalline powder with about 500 times the sweetening power of sucrose. It is not very soluble in water but is used extensively as a sweetening agent, in the form of the sodium salt.

salt a compound formed when an ACID reacts with a BASE or when a metal atom replaces one or more hydrogen atoms of an acid. Salts can be formed by other reactions such as two salts reacting together to form different compounds. Salt is also the name given to common salt, or sodium chloride (NaCl), which is found in sea water. It is a vital component of our diet although too much is considered harmful.

sandstone a SEDIMENTARY ROCK made up of sand grains with sizes between 0.06 mm and 1 mm, and a variety of other minerals and materials cementing the grains together. CALCITE is a common cement, and silica (as QUARTZ) cements sands to produce a hard sandstone often referred to as an orthoquartzite (to tell it apart from the metamorphic rock, quartzite). Brown and red sandstones usually have an iron-rich cement such as LIMONITE or haematite. Other minerals which may be present include FELDSPAR and MICA.

saponification a process in which ESTERS are hydrolysed (see HYDROLYSIS) by the action of acids, alkalis, boiling with water or superheated steam. If alkalis are used then SOAPS are produced and this is the origin of the term (*sapo* is the Latin for soap).

saprotroph an organism that obtains its nutrition from dead and decaying organic matter. The group includes many bacteria and fungi, which are responsible for the release of nitrogen, carbon dioxide and other nutrients from the decomposing matter.

satellite any body, whether natural or manmade, that orbits a much larger body under the force of gravitation. Hence the MOON is a *natural satellite* of the Earth. All the planets, except for MERCURY and VENUS, have at least one natural satellite.

Artificial satellites are manmade spacecraft launched into orbit from Earth. The first satellite to be launched was the Russian *Sputnik I* in 1957, but many hundreds have followed since then. Some satellites, especially those used for communications, are placed in a special geostationary orbit. This orbit is about 36,000 km above the Earth's surface. The satellite orbits the Earth in the same period of time as Earth rotates on its axis (24 hours). Hence the satellite maintains the same position relative to the Earth and appears to be stationary. Equipment on Earth therefore does not need to be adjusted to follow the satellite.

Many other satellites are used for other purposes such as meteorological recordings and weather forecasting. Each satellite has a dish aerial facing towards Earth and *thruster* motors to help maintain its position. When the fuel supply for the thrusters is exhausted, the satellite drifts out of its orbit and can no longer be used. The equipment on board a satellite or spaceship requires electricity which is usually derived from solar powered cells. (However, spacecraft travelling long distances away from the Sun have electricity generated by small nuclear reactors).

saturated compound a group of organic compounds with no double or triple BONDS; i.e., structurally they are full and do not form ADDITION compounds through the joining of hydrogen atoms or their equivalent.

saturated solution a SOLUTION of a substance (solute) that exists in EQUILIBRIUM with excess SOLUTE present. Heating a saturated solution allows more solute to dissolve to form a supersaturated solution. Cooling of the solution or a loss of solvent will cause some of the solute to come out of solution, that is, to crystallize.

Saturn one of the four *gas giants*, the second largest planet and sixth in the SOLAR SYSTEM, with an orbit between that of JUPITER and URANUS. Saturn has a diameter at the equator of about 120,800 kilometres and is a maximum distance of 1,507,000,000 kilometres from the SUN. Saturn rotates very fast and this causes it to flatten at its poles and bulge at the equator. A day on Saturn lasts for $10^1/_4$ hours and a year, (or one complete orbit of the Sun), for 29.45 Earth years. Saturn is a cold planet of frozen gases and ice and a surface temperature in the region of -170°C. It is mainly gaseous with an outer zone of *hydrogen* and *helium* over a metallic hydrogen layer and a core of ice silicate. The atmosphere is rich in methane and ethane. Saturn is well-known for its *rings* which are, in fact, ice particles and other debris thought to be the remains of a SATELLITE which broke up close to the planet. The rings are wide, in the region of 267,876 kilometres across, but they are extremely thin (only a few kilometres). The *Voyager* space probes approached close to Saturn in 1980 and 1981 and photographs taken revealed that there were many more rings than had previously been detected. They are brighter than those of any other planet.

Saturn has 24 satellites or moons, some of which were discovered by the *Voyager* spacecraft, including *Atlas, Prometheus* and *Calypso. Titan* is the largest moon and, at 5200 kilometres diameter, is larger than MERCURY. It is the only moon known to have a detectable atmosphere, a layer of gases above its surface.

savanna areas of grassland in tropical or sub-tropical zones, occupying a broad zone between the tropical forests and the semi-arid steppes. They are a result of prolonged lack of water in soils and usually occur in areas of low relief where rainfall due to mountains is absent.

scanning electron microscope see **electron microscope**.

scattering the dispersal of waves or particles upon impact with matter; applicable to light, atomic particles, etc.

schist a PELITIC ROCK (that is made up of clay minerals) formed by REGIONAL METAMORPHISM. Schist is the higher metamorphic grade, coarse-grained equivalent of PHYLLITE and displays a SCHISTOSITY. The minerals present vary but will usually include MICA (biotite or muscovite), QUARTZ, FELDSPAR, often GARNET and many other accessory minerals.

schistosity the planar FABRIC (foliation) formed in a schist due to the alignment of minerals, predominantly MICAS, but also AMPHIBOLES. The alignment is generated either by the physical rotation of minerals during deformation or the metamorphic growth of new minerals which preferentially align themselves in the fabric which is at right angles to the direction of maximum compression.

Schottky defect refers to the less than ideal arrangement of atoms or ions in a crystal structure in which an atom or ion is completely missing. Defects such as this may produce anomalous physical properties e.g. colour, conductivity. These properties can be utilized in catalysis, semiconductors and elsewhere.

Schrödinger (wave) equation the basic equation used in wave mechanics which describes the behaviour of a particle in a force field.

scientific notation a useful method of writing large and small numbers. The scientific notation for a number is that number written as a power of 10 times another number, x, such that x is between 1 and 10 (1 < x <10), e.g. 145,800 = 1.458×10^5.

scintillation small light flashes caused by ionizing radiations (α, β, or γ rays) striking substances that luminesce.

In astronomy, scintillation is the twinkling of stars. The light rays from stars are effectively from point sources, and the twinkling is due to deflection by the Earth's atmosphere.

scleroproteins insoluble PROTEINS which form the skeletal parts of tissues. Included are KERATIN, COLLAGEN and elastin (a fibrous protein found in lungs, artery walls and ligaments).

scree the accumulation of mainly coarse, angular rock debris at the foot of cliffs inland. The debris is produced by the weathering and gradual disintegration of the upper slopes and cliffs, through the agencies of frost and water.

screw is one of the six varieties of simple *machine*. It is an extremely effective device and is used extensively for fastening things together. Screws or screw threads are also used in instruments to enable controlled movement, e.g. on a microscope stage, or in a micrometer when making accurate measurements. A screw jack is one type of jack for lifting a car to allow a wheel to be changed. This means that a heavy object like a car can be raised relatively easily.

sea-floor spreading (*see also* MID-OCEANIC RIDGE) the theory, which contributed greatly to the overall concept of PLATE TECTONICS, that ocean floor is created at ridges running along the centres of the ocean floors and that these are the margins of new tectonic plates. MAGMA of a basaltic composition rises along the ridge and the newly generated crust spreads away from the ridge.

sea-level pressure the value of atmospheric pressure at sea level calculated from the pressure at the measuring point and the height to sea-level. The value is used on meteorological charts.

seat earth a fossil soil (palaeosol) which is found immediately beneath coal seams and represents the soil in which the vegetation grew. It is the last sediment deposited before plant life became established and frequently contains fossilized roots.

sea water all but 0.1% of material dissolved in sea water is due to eleven components (see below). The bulk of the calcium and bicarbonate ions precipitate out as CALCIUM CARBONATE, silica is taken up by organisms and most of the material dissolved consists of five ions; chloride, sodium, sulphate, magnesium and potassium.

ion	‰ (parts per thousand)
chloride Cl⁻	19.0
sodium Na⁺	10.6
sulphate SO_4^{2-}	2.6
magnesium Mg^{2+}	1.3
calcium Ca^{2+}	0.4
potassium K⁺	0.4
bicarbonate HCO_3^-	0.1
bromide Br⁻	
borate $H_3BO_3^-$	less than 0.1‰
strontium Sr^{2+}	
fluoride F⁻	

secant the function of an angle in a right-angled triangle, given by the reciprocal of the COSINE function: the secant of an angle A is 1/cosA.

second a unit of plane angle, equal to 1/60th of a minute or 1/3,600th of a degree, or $\pi/648,000$ radian. Also 1/86400th of the mean solar day.

secondary metabolite a general term applied to several groups of compounds that are not directly involved with the biological processes that contribute to growth, such as photosynthesis and respiration. However, they may be chemicals used in defence or similar mechanisms. Typical groups are TERPENOIDS, alkaloids and the flavonoids (plant pigments).

sedimentary rock are rocks formed from existing rock sources through the processes of erosion, weathering and include rocks of organic or chemical origin. They can be divided into *clastic* rocks, that is made of fragments, *organic* or *chemical*.

Typical Sedimentary Rocks		
Clastic	**Organic**	**Chemical**
sandstone	LIMESTONE	limestone
shale	ironstone	dolomite
mudstone	chert	flint
conglomerate	coals	gypsum and other EVAPORITES
limestone		

The clastic rocks are further divided on grain size into coarse (or *rudaceous*, grains of 1-2 mm), medium (or *arenaceous*, eg sandstone) and fine (or *argillaceous* up to 0.06 mm). When the grains comprising clastic rocks are deposited (usually in water) compaction of the soft sediment and subsequent *lithification* (that is, turning into rock) produces the layered effect, or *bedding*, that is often visible in cliffs and outcrops in rivers. It is also common for original features to be preserved, for example ripples, small or large dune structures, which in an exposed rock face appear as inclined beds called *current bedding*. *Graded bedding* shows a gradual change in grain size from the base, where it is coarse, to the top of a bed, where it is fine and this is due to the settling of material onto the sea floor from a current caused by some earth movement.

Many sedimentary rocks, particularly shale, limestone and finer sandstone contain FOSSILS of animals and plants from millions of years ago, and with the original features mentioned above, these are useful in working out the sequence of events in an area where the rocks have been strongly folded.

sedimentary structure a fossil feature, preserved in or on the bedding of a sedimentary rock. The structures are those made by sedimentary processes or the activity of organisms at the time of deposition of the sediments. Structures preserved on bedding surfaces include RIPPLE MARKS, scour marks (caused by erosion) and tool marks (caused by an object being carried over the surface), and these are all formed by depositional processes. SOLE MARKS are structures preserved on the bases of beds and include trails, tool marks and infilled scours and load casts (a bulging formed at the base of a bed where the upper bed sinks into the lower while the sediment is wet—forming a lobe). Internal sedimentary structures include those formed by depositional processes e.g., laminations, convolute bedding (a distorted bedding due to expulsion of water from sediments deposited quickly) and slump (overturned folds due to sliding sediment) structures; organic activity e.g., bioturbation or by chemical activity after deposition e.g., concretions (often an egg-shaped lump of a mineral caused by early hardening within a sediment).

seismograph in the study of earthquakes (*seismology*), seismographs record the shock waves (*seismic* waves) as they spread out from the source. The seismograph has some means of conducting the ground vibrations through a device to turn movement into a signal that can be recorded. There are numerous seismic stations around the world that record ground movements, each contains several seismographs with *seismometers* (the actual detector linked to a seismograph).

seismology the study of EARTHQUAKES and their shock (seismic) waves which helps scientists to achieve a greater understanding of the deep structure of the Earth. Exploration seismology uses explosive charges to generate waves to determine structure and search for mineral resources e.g., hydrocarbons.

semicircular canals a structure in the inner ear in vertebrates which helps maintain dynamic equilibrium (balance). It comprises three 'loops' at right angles to each other and in each loop is a fluid (the endolymph). When the head is moved, the fluid moves accordingly and sensory cells respond to the movement of the endolymph. The nerve impulses are then transmitted to the brain.

semiconductor an element or compound with average resistivity (RESISTANCE related to dimensions) between that of a CONDUCTOR and an INSULATOR. The commonest are silicon, germanium, selenium, and lead sulphide. These and other materials form the basis of TRANSISTORS, DIODES, thyristors, and integrated circuits. The transistor is the basis of electronic circuits and consists of minute slices of silicon in a sandwich, altered chemically to confer differing conductivities. The passage of current through the sandwich can be controlled by a weak current through the central slice, thus creating an electrically operated switch. The addition of impurities to semiconductors also changes their properties. This is called doping, and is a technique used to make p-type and n-type components in semiconductors which are used in electronic circuits.

senescence the process of ageing that is characterized by the progressive deterioration of tissues and the metabolic functions of their cells. According to research, senescence may be caused by the accumulation of genetic mutations within the body's cells or the expression of undesirable GENES in the later years of an individual's life. Some organisms are able to suppress senescence by REGENERATION, a process common in many simple invertebrates and one achieved by some plants using VEGETATIVE PROPAGATION.

sequence a succession of mathematical entities, $x_1, x_2, ..., x_n, ...$, which is indexed by the positive integers. The term x_n is the nth term or general term. A sequence may be defined by stating its nth term, e.g. a sequence whose nth term is $n/n + 1$ is the sequence $1/2, 2/3, 3/4, 4/5, 5/6, ..., n/n + 1, ...$

series an expansion of the form $x_1 + x_2 + x_3 + ..., + x_n + ...$, where xn are real or complex numbers.

In geology, series is a major part of a SYSTEM, referring to the rock formed during one EPOCH (see *also* APPENDIX 5). A series can be subdivided into stages.

series circuit where a common current flows through the components in a circuit.

serum see **blood serum**.

sex chromosome one of a pair of chromosomes that play a major role in determining the sex of the bearer, with a different combination in either sex. An individual is said to be homogametic when it has a HOMOLOGOUS pair of sex CHROMOSOMES (as in the XX of female mammals) and is said to be heterogametic when it has different sex chromosomes forming its pair (as in the XY of male mammals). Sex chromosomes contain GENES that decide an individual's sex by controlling the sexual characteristics of the individual, e.g. testes in human males and ovaries, breasts, etc in human females.

sexual reproduction the production of progeny that have initially arisen from the fusion of male and female GAMETES in a process called FERTILIZATION. In DIPLOID organisms, sexual reproduction must be preceded by MEIOSIS to form the HAPLOID gametes if there is not to be a doubling of the number of chromosomes in all sexually reproduced offspring. As sexual reproduction involves MEIOSIS, it introduces greater genetic variation in a species, because GENETIC RECOMBINATION can occur during meiosis, with the result that any offspring will have gene combinations that differ from its parents.

Seyfert galaxy a small category of GALAXY which exhibits a bright nucleus (similar in many ways to a QUASAR) and weak spiral arms. They emit infrared radiation and, to a lesser extent, radio waves and X-rays (see ELECTROMAGNETIC WAVES).

shale a sedimentary rock which is fine-grained (composed of silt, mud and clay-sized particles). Shales are fissile (split easily) due to the alignment of clay and similar minerals with their flat surfaces parallel to the planar FABRIC.

shard a volcanic glass fragment found in certain PYROCLASTIC rocks.

shear zone in geology, a zone which is usually narrow compared to its length, in which there is intense deformation due to the relative movement of two adjacent undeformed blocks of rock. There is a continuous range of deformation from brittle (represented by FAULTS), to ductile (as in FOLDS etc., where there is more 'flow'). Brittle faulting occurs in the upper portion of the crust (at depths of 10 to 15 km) and below this level ductile flow dominates, so that the surface expression of ductile flow at depth will be faulting. The variation in strain across a shear zone produces a progressive development in the fabric, compared to the undeformed rock. Mineral grains change shape, becoming more elongated nearer the centre of the zone and at the same time rotating to become parallel to the shear direction where deformation is most intense.

shell star a star surrounded by a layer of luminous gas. Such stars belong to SPECTRAL TYPES O, B or A.

shield a screen placed around persons or equipment to offer protection against harmful rays (e.g., X-rays) or to protect an electronic component from interference from electromagnetic fields. In geology, a stable area of the crust, also called a craton.

short-period comet a COMET with an orbit inside the SOLAR SYSTEM and a period of under 150 years e.g., Halley's Comet.

sidereal day the time taken for the Earth to make one complete rotation upon its axis with reference to the fixed stars. This is a little over four minutes shorter than the MEAN SOLAR DAY.

sidereal month the time taken by the moon in completing one orbit of the Earth, determined using a fixed star as a reference point. It is 27 days, 7 hours, 43 minutes.

sidereal period the periods taken by the Moon and planets to reach the identical successive positions, relative to the line joining the Earth to the Sun.

sidereal time the measurement of time using the rotation of the EARTH with reference to distant stars (and not the Sun as with 'civil' time).

sidereal year the period in which the SUN appears to make one revolution with reference to the fixed stars. It is 365.26 MEAN SOLAR DAYS, slightly longer than the civil year (365.24).

sievert the SI unit of radiation dose defined as that radiation delivered in one hour at a distance of one centimetre from a point source of one milligram of radium enclosed in platinum which is 0.5mm thick.

sigma the symbol Σ (Greek capital Sigma—"S" for "sum") used in statistics and mathematics. For example, Σx means the sum of all the values that x can assume, and Σx^2 means square each value of x, then add the results.

sigma bond (σ) the bond type formed in a carbon-carbon single bond where two atomic ORBITALS overlap to form a molecular orbital surrounding the two carbon nuclei.

significant figures the digits in a number which contribute to its value e.g. in the number 0.762 the zero is insignificant whereas the other digits are significant. If the number were 0.7620, the last zero ought to be significant because it should indicate that the number is accurate to four decimal places. However, the final zero as shown here is often added arbitrarily by the originator of the data and it may not represent such accuracy.

silica silicon dioxide (SiO_2), is one of the most important constituents of the Earth's crust. It is polymorphic (i.e., exists in more than one crystal form—when referring to elements it is called *allotropy*) with QUARTZ existing up to

573°C, then tridymite (to 1470°C) and finally cristobalite (to 1710°C, the melting point). High pressure forms are also known (coesite, keatite and stishovite) but these are mainly experimental products. In addition to the mineral form quartz, silica also occurs in cryptocrystalline (very finely crystalline) forms (chert, FLINT, chalcedony) and as amorphous opal (the hydrated form).

Silica is used in the manufacture of glass and refractories and the latter in the form of gannister (> 90% SiO_2) is used in furnace hearths.

silica gel the hard AMORPHOUS form of hydrated silica, which is chemically inert, but highly HYGROSCOPIC. It is used for absorbing water and solvent vapours and other drying or refining tasks and can be regenerated by heating.

silicates an enormous group of rock-forming minerals ranging from the simple zircon, $ZrSiO_4$ which has discrete $(SiO_4)^{4+}$ anions, to highly complex structures. All are based on the SiO_4 group in a tetrahedral (like a pyramid on a triangular base) formation with some sharing of the oxygen atoms. Aluminium, beryllium and boron can replace silica in the $(SiO_4)^{4+}$ ANION, and the CATIONS occupy the holes in the lattice. Silicate minerals are classified on the arrangement of the SiO_4 tetrahedra as follows:

neosilicates	SiO_4 tetrahedra linked by cations e.g. zircon and OLIVINE
sorosilicates	two tetrahedra sharing one oxygen e.g. epidote group
cyclosilicates	rings of 3, 4 or 6 SiO_4 sharing two oxygens e.g. BERYL
inosilicates	single chains with 2 linking oxygens e.g. PYROXENE or double chains with 2 or 3 linking oxygens e.g. AMPHIBOLES
phyllosilicates	sheets in hexagonal networks, sharing 3 oxygens e.g. MICAS
tectosilicates	frameworks in three dimensions e.g. QUARTZ and FELDSPARS.

siliceous deposits any deposit with a high silica content, whether deposited chemically, mechanically or due to accumulation of skeletal remains (e.g. of diatoms).

silicon is a non-metallic element and the second most abundant in the Earth's crust. It does not occur as free silicon, but is found in abundance as numerous silicate minerals including quartz (SiO_2). Silicon is manufactured by reducing SiO_2 in an electric furnace but further processing is necessary to obtain pure silicon. When doped with boron or phosphorus it is used in SEMICONDUCTORS, (see also DIODE). Quartz and some silicates are used industrially to produce glass and building materials. Silicon melts at 1410°C.

silicon chip a SEMICONDUCTOR chip of crystalline silicon onto which is printed a microelectronic (integrated) circuit for use in computers, radios, etc.

silicones organic polymers with silicon built on SiR_2O groups. Because they are inert, colourless and odourless, their uses are numerous and the properties depend upon the degree of polymerization. They are used in oils and greases, sealing compounds, resins (for insulation and laminates). The simpler compounds form oils while the more complex varieties form good electrical insulators.

sill an igneous INTRUSION which pushes into the surrounding rocks, and often takes advantage of weaknesses by bedding planes. In such cases, sills therefore lie parallel to the bedding direction of the rocks. Sills show a range of sizes and large ones may exceed 100 metres in thickness.

simple harmonic motion motion which is characteristic of many systems that vibrate or oscillate. If a point moves around a circle with a constant angular velocity, the projection of this point onto the circle's diameter (by means of a perpendicular line) produces another point which moves back and forth along the diameter as the point moves around the circle. This is a basic illustration of simple harmonic motion. To project this moving point along an axis would produce a sine wave. Typical examples are a mass 'bouncing' on a spring, a swinging pendulum and the oscillations of air as a sound wave passes.

simultaneous equations two or more equations with two or more unknown variables, which may have a unique solution, e.g., $4a - b = 10$ and $3a + 2b = 24$ yields the solution $a = 4$ and $b = 6$.

sine a function of an angle in a right-angled triangle, defined as the ratio of the length of the side opposite the angle to the length of the HYPOTENUSE.

sine rule in any triangle $A/\sin a = B/\sin b = C/\sin c$; or, any side divided by the sine of the opposite angle is equal to any other side divided by the sine of its opposite angle.

single bond see **sigma bond**.

SI units a system of coherent metric units—Système Internationale d'Unités (see APPENDIX 4).

skin an important organ which forms the outer covering of a vertebrate animal. There may be a variety of structures protruding from the surface of the skin (i.e., hair, fur, feathers or scales) depending upon the type of animal. The skin provides a protective layer for the body, helping to cushion it in the event of accidental knocks and preventing drying out. Also, skin helps to maintain the correct body temperature. When the body is hot the many blood *capillaries* in the skin widen allowing more blood to flow through them, and heat is lost to the outside by radiation. Sweat glands present in the skin secrete a salty fluid which evaporates from the surface and forms another cooling mechanism. If the body is cold, the capillaries contract to decrease blood flow and conserve heat. The layer of fat in the skin has a warming effect and small muscles attached to hair roots contract to raise the hair. The hairs trap a layer of air which helps to warm the body. Skin is also a physical barrier to harmful substances or organisms which might otherwise enter more easily.

slate a low grade fine-grained METAMORPHIC rock. Due to the intense compression during deformation of rocks, sheet SILICATE minerals (e.g. MICAS, chlorite) become aligned, imparting a fissility to the rock known as slaty CLEAVAGE. Slate is used commercially for roofing in particular, but also larger slabs are used for other items such as billiard-table tops.

slumping the process whereby rocks or sediments slide downslope under the influence of gravity. It is a common occurrence in subaqueous environments with sediments that are not completely consolidated. On land, slumping occurs on cuttings, undercut slopes and cliffs where there is a slip surface to facilitate movement and no obstacle in front of the slipping rock and/or soil. It occurs on spoon-shaped shear planes which leave an arc-like trace on the ground.

snow precipitation of small ice crystals falling singly or in flakes. The crystals form from water vapour in cloud.

soap the sodium or potassium salts of the FATTY ACIDS, stearic, palmitic and oleic acid. Soaps are produced by the action of sodium or potassium hydroxide on fats (see SAPONIFICATION), and is made by heating fat (or a vegetable oil) with sodium hydroxide to which salt is added. Perfumes may then be added before it is shaped into tablets.

sodium (Na) is an alkali metal which occurs widely—its principal source being sodium chloride (salt) in sea water and salt deposits. It is obtained by electrolysis of fused sodium chloride, but does not exist in its elemental form because it is highly reactive. When prepared, the metal is sufficiently soft to be cut with a knife and it is a silvery white colour. However, it reacts violently with water and rapidly with oxygen and the halogens. It is essential to life, particularly in the biological mechanism involved in the transmission of nerve impulses.

It forms numerous compounds with diverse uses and is itself used as a heat-transfer fluid in reactors. Compounds and their uses include: hydroxide, numerous uses: benzoate, antiseptic; carbonate, glass, soap and other manufacturing processes; chlorate, herbicide; citrate, medicinal; hypochlorite, bleaches; nitrite, in dyes, and many more.

sodium chloride NaCl (see HALITE) or rock salt. A white crystalline solid obtained either from halite or the evaporation of sea water. It is used in the production of CHLORINE and in the alkali and glass industries and is one of the most important raw materials for the chemical industries.

sodium hydroxide caustic soda, NaOH, a whitish substance that gives a strongly alkaline (see ALKALIS) solution in water and will burn skin. It is *deliquescent*, that is it picks up moisture from the air and may eventually liquefy. It is used a great deal in the laboratory and is a very important industrial chemical, being used in the manufacture of pulp and paper, soap and detergents, petrochemicals, textiles and other chemicals. Sodium hydroxide is itself manufactured by the electrolysis of sodium chloride solution, chlorine being the other product, or by the addition of hot sodium carbonate solution to quicklime (calcium oxide, CaO).

solar constant the sum total of the electromagnetic energy from the Sun, measured over the Earth's mean distance per unit time (currently $1.37kWm^{-2}$).

In physics, it is the energy *received* on the Earth's surface allowing for any losses due to the atmosphere.

solar corona the outer layer of the Sun's atmosphere which reaches temperatures of two million K. It is a strong source of X-rays and is visible as the halo of light during an eclipse.

solar energy is energy which reaches Earth from the sun in the form of heat and light. 85% of solar energy is reflected back into space and only 15% reaches the Earth's surface, but it is this energy which sustains all life.

solar flare a temporary eruption of solar material from the Sun's surface generating intense radio emissions and particles with very high energies. Solar flares are driven by magnetic forces.

solar system the system comprising the Sun (a star of SPECTRAL TYPE G) around which are the nine planets in elliptical orbits. Nearest the Sun is MERCURY, then VENUS, EARTH, MARS, JUPITER, SATURN, URANUS, NEPTUNE and PLUTO. In addition there are numerous satellites, a few thousand (discovered) asteroids and millions of comets. The age of the solar system is put at 4.5 to 4.6 billion years, a figure determined by the RADIOMETRIC DATING (uranium-lead) of IRON METEORITES. Iron meteorites are thought to be fragments of cores from early planets and thus representative of the early stages of the solar system.

solar wind the term for the stream of charged, high-energy particles (primarily ELECTRONS, PROTONS and alpha particles) emitted by the sun. The particles travel at hundreds of kilometres per second and the wind is greatest during flare and sunspot activity. Around the EARTH the particles have velocities of 300-500 kms^{-1}

and some become trapped in the magnetic field to form the VAN ALLEN RADIATION BELT. However, some reach the upper atmosphere and move to the poles producing the auroral displays (see AURORA).

sole mark see **sole structure**.

sole structure in geology SEDIMENTARY STRUCTURES found on the base of beds formed primarily by the scouring of a current or the movement of an object (called a tool) over the sediment surface. The impression created is then often filled by sands and other sediment. Other features include flutes (tongue shapes scoured out of the mud) created by the turbulent water and a variety of shapes caused by the passage of an object i.e., whether it is dragged, bounced, etc. These structures help to determine the way-up of beds and can assist in derivation of the palaeocurrent, that is the direction of the current when the sediments were deposited millions of years ago.

solenoid a tightly wound, cylindrical coil of wire, which generates a magnetic field when current is passed through the coil.

solid in geometry, a figure with the dimensions of length, width and breadth and thus a measurable VOLUME. In chemistry, a state of matter in which the component MOLECULES, ATOMS, or IONS hold a constant position in relation to one another, i.e., they exhibit no translational motion. Even though solids are rigid the atoms can vibrate, and if heat is applied the vibrations increase until the structure breaks up and it melts to form a liquid.

solid state physics the study of all the properties of solid materials, but especially of SEMICONDUCTORS and "solid-state" devices, i.e., devices with no moving parts, as in integrated circuits, TRANSISTORS, etc.

solstice the time at which the SUN reaches its most extreme position north or south of the equator. There are two such instants in the year.

solubility the concentration of a SATURATED SOLUTION is called the solubility of the given solute in the particular solvent used, measured in kgm^{-3}.

solubility product the number of IONS of (each type in) a compound in solution that can exist together, i.e., the product of the concentration of the ions when in equilibrium (which, for a slightly soluble salt, is a constant at a set temperature). When the solubility product is exceeded in a solution, the compound (i.e. both ions after they have recombined) will be precipitated until the product falls to the constant value.

solute one substance dissolved in another. A solute dissolves in a SOLVENT to form a SOLUTION.

solution a single mixture of two or more components in one phase, which usually applies to solids in liquids and often refers to a solution in water (aqueous solution). However, other solutions include gases in liquids and liquids in liquids.

solvent a substance, usually a liquid, that can dissolve or form a SOLUTION with another substance.

somatic cell any of the cells of a multicellular organism (plant or animal) other than the reproductive cells (GAMETES).

sonar (*acronym for* Sound Navigation Ranging) a device that transmits high frequency sound and collects returning sound waves that have been reflected from submerged objects. The depth is indicated by the time taken for the return journey.

sonic boom the loud bang created by shock waves from the leading and trailing edges of an aircraft travelling supersonically. When an aircraft increases its speed near to the speed of sound, a wave of compressed air (shock waves) forms in front of it. The boom results from the aircraft overtaking the pressure waves it creates ahead of itself.

sound the effect upon the ear created by air vibrations with a frequency between 20 Hz (hertz) and 20 kKz (20,000 Hz). More generally, sound waves are caused by vibrations through a medium (whether gas, liquid or solid). One of the commonest sources of sound is a loudspeaker. When it produces sound, the cone vibrates, producing a series of compressions in the air. These are called longitudinal progressive waves, that is the oscillations occur in the same direction that the wave is travelling. The sound waves so produced enter the ear causing pressure changes on the ear drum, causing the brain to register the sound. Most items produce sound when they vibrate or are moved or banged together, but sound can only be transmitted through a medium and it cannot travel through a vacuum.

The speed of sound varies with the material it is travelling through, moving most quickly through solids, then liquids and gases. In air, sound travels at approximately 350 m/s. The speed increases with temperature, but is unaffected by pressure. Frequency of sound waves relates to *pitch*; high frequencies produce a sound of high pitch, e.g. a whistle at 10 kHz, while low pitch is caused by low frequencies, e.g. a bass voice at 100 Hz. Sound intensity is measured in *decibels* (db) and it is a logarithmic scale. Ordinary conversation might register 40-50 decibels, traffic 80 and thunder 100 while jet aircraft can exceed 125 dB.

Southern Cross a constellation of the southern hemisphere comprising a cross of four stars, identified by its position to the west of two bright stars α and β Centauri.

Southern Lights see **aurora**.

space velocity a star's movement and direction in three-dimensional space.

species a group of individuals that can potentially or actually breed among themselves producing viable offspring, and that within the group may show gradual morphological variations but remain different to other groups. In the taxonomic classification, species are grouped into a genus (plural *genera*) and species can themselves be subdivided into subspecies, varieties, etc. The naming of species is governed by the system of *binomial nomenclature*, so that a generic and specific name (in Latin) identify a particular individual, e.g. *Panthera pardus* is the leopard.

specific heat capacity the heat required by unit mass to raise its temperature by one degree (SI units—joules per kg per Kelvin).

specific humidity (see *also* HUMIDITY) the mass of water vapour in air in proportion to the total mass of the air.

specific latent heat see **latent heat**.

speckle interferometry a technique for measuring small angles e.g. the diameter of stars, which utilises the principle of interference of light.

spectral types a classification system for stars, based upon the spectrum of light they emit. The sequence is, in order of descending temperature: O - hottest blue stars; B - hot blue stars; A - blue white stars; F - white stars; G - yellow stars; K - orange stars; M - coolest red stars. The system originates from Harvard College Observatory which originally had classes A to Q but this was altered and ordered by temperature to give the quoted classes which can be further subdivided into categories numbered 0 to 9. There are also more recent additions such as S stars and carbon stars. The temperature ranges are as follows:

type		
	O	>25 000K
	B	11 000 - 25 000K
	A	7500 - 11 000K
	F	6000 - 7500K
	G	5000 - 6000K
	K	3500 - 5000K
	M	<3500K

In addition, a number of descriptive letters are used as prefixes and suffixes to provide further information about the stars' spectra, including:

c	sharp lines
d	dwarf (main sequence star)
D	white dwarf
e	emission
ep	peculiar emission
g	giant
k	interstellarlines
m	strong metallic lines
n	diffuse lines
s	sharp lines
sd	subdwarf
wd	white dwarf
wk	weak lines

The luminosity of stars was also subdivided:

Ia	luminous supergiants
Ib	less luminous supergiants
II	bright giants
III	normal giants
IV	subgiants
V	dwarfs (main-sequence stars).

This enables any star to be classified readily.

spectrochemical analysis the heating of a sample to a high temperature producing emission lines which relate to abundance of the elements in the sample.

spectroheliograph an instrument for photographing the Sun using a single wavelength of light (i.e. monochromatic light).

spectrohelioscope a device which is essentially the same as a SPECTROHELIOGRAPH but which permits an image of the Sun to be viewed in light of one wavelength.

spectroscope the general term for the equipment used in spectroscopy. It consists basically of a slit from which a beam of radiation emerges, a collimator which makes the beam parallel, the prism which disperses the varying wavelengths and some device for counting/measuring the radiation.

spectrum the separation of a beam of electromagnetic radiation into its constituents which are defined by their different wavelengths. An obvious example is the coloured bands obtained when white light passes through a prism, or the natural equivalent which is the rainbow. The full spectrum reaches from radio waves to gamma radiation and includes visible light, microwaves and X-rays.

speed for a body moving in a straight line or continuous curve, the ratio of distance covered to the time required to cover that distance. Units vary, e.g. metres per second (ms^{-1}), miles per hour (mph).

speed of light as revealed in the theory of RELATIVITY, a universal and absolute (that is, independent of the speed of the observer) value that is 2.998×10^8 ms^{-1}, or 186281 miles per second.

speed of sound the value for the speed (VELOCITY) of sound depends upon the nature of the medium and the temperature. In air at 0°C, the speed is 332 ms^{-1}, or about 760 mph. In fresh water, the speed is 1410 ms^{-1}.

sphere a circular solid figure with all points on its surface an equal distance from the centre. In two-dimensional CARTESIAN CO-ORDINATES, the equation of a sphere is $(x - a)^2 + (y - b)^2 + (z - c)^2 = r^2$. For a sphere of radius r, vol = $4/3\pi r^3$; surface area A = $4\pi r^2$.

spiral galaxy a galaxy which has spiral arms containing the stars, dust and gas. The SOLAR SYSTEM belongs to such a galaxy. Young stars are concentrated in the arms and older stars and globular clusters occur in a halo type of structure surrounding the arms.

spontaneous combustion the ignition of a substance without application of a flame. It may occur through the production of heat from slow OXIDATION within the substance.

spontaneous generation a theory once held that now has no credit, that living organisms could arise from non-life. It was believed that spontaneous generation could occur in, for example, rotting meat or fermenting broth, giving rise to an individual organism, but it is now known that all new organisms originate from the parent organism from whom they have inherited a genetic ancestry.

spore a small reproductive unit, usually consisting of one cell, that detaches from the parent and disperses to give rise to a new individual under favourable environmental conditions. Spores are particularly common in fungi and bacteria but also occur in all groups of green land plants such as ferns, horsetails and mosses.

square see **polygon**.

stalactite a hanging deposit of calcium carbonate ($CaCO_3$) formed from the roof of a cave by drips of calcium-rich solutions. Stalactites resemble icicles.

stalagmite an upstanding growth of calcium carbonate ($CaCO_3$) formed on the floor of a cave by drips of calcium-rich solutions. Often found with stalactites, when the two forms may eventually meet and join. The solutions form due to groundwater running through limestone rock and because there is little evaporation from the drops of water in underground caves, some of the dissolved material is precipitated.

standard time the time in the time zones established by international agreement. Each zone is equal to one hour and is approximately 15° of longitude wide.

standing wave a disturbance produced when two similar wave motions are transmitted in opposite directions at the same time. This results in INTERFERENCE, with the combined wave effects producing maxima and minima over the area of interference. The resultant waveform is contained within fixed points and does not move, hence standing or stationary wave.

star a body of matter, similar to the SUN, which is contained by its own gravitational field. Stars are glowing masses that produce energy by thermonuclear reactions (NUCLEAR FUSION). The core acts as a natural nuclear reactor where HYDROGEN is consumed and forms HELIUM with the production of electromagnetic radiation (see ELECTROMAGNETIC WAVES). Stars can be classified by the spectrum of the light they emit (see SPECTRAL TYPES).

starch is a polysaccharide (see SACCHARIDE) found in all green plants. Starch is built up from chains of glucose ($C_6H_{12}O_6$) units arranged in two ways, as amylose (long, unbranched chains) and amylopectin (long, cross-linked chains). Potato and some cereal starches contain 20-30% amylose and 70-80% amylopectin. Amylose contains 200-1000 glucose units while amylopectin numbers about 20. Starch is formed and broken down in plant cells and is stored as granules and it occurs in seeds. It is insoluble in cold water and is obtained from corn, wheat, potatoes, rice and other cereals by various physical processes. It is used as an adhesive for sizing paper and has many uses in the food industry.

static electricity electric charges at rest which result from the electrostatic field produced by the charge (see *also* VAN DER GRAAF GENERATOR). In some materials, some of the electrons that orbit in a cloud around the nucleus can be pulled off if the material is rubbed with a cloth, leaving the material charged.

statics a branch of mechanics which studies the combination of forces in equilibrium, i.e., the behaviour of matter under applied forces when there is no resulting motion.

stationary orbit is the orbit of a satellite around a body such that the satellite stays above a fixed point about the latter's equator. For the earth the GEOSTATIONARY orbit is approximately 36,000 km above the equator. This is a property used in the positioning of communications satellites.

statistics the branch of mathematical science dealing with the collection, analysis and presentation of quantitative data, and drawing inferences from data samples by the use of probability theory.

steel is iron that contains up to 1.5% carbon in the form of *cementite* (Fe_3C). The properties of steel vary with iron content and also depend upon the presence of other metals and the production method. *Alloy steels* contain alloying elements while *austenitic steel* is a solid solution of carbon in a form of iron and is normally stable only at high temperatures but can be produced by rapid cooling. *Stainless steel* is a group of chromium/nickel steels which have a high resistance to corrosion and chemical attack. A high proportion of chromium is necessary (12-25%) to provide the resistance and a low carbon content, typically 0.1%. Stainless steel has many uses: cutlery, equipment in chemical plants; ball bearings and many other items of machinery.

stellar evolution the various stages of the life of a star which begins with the creation of the star from the condensation of gas, primarily hydrogen. The growth of the clouds pulls in more gas and the increase in gravity compresses the molecules together which attracts more material and creates a denser mass. The heat normally produced by molecules due to their vibratory motion is increased greatly and the temperature is raised to millions of degrees which facilitates NUCLEAR FUSION. The supply of hydrogen continues to be consumed (and the star occupies the MAIN SEQUENCE of the HERTZSPRUNG-RUSSELL DIAGRAM) until about 10% has gone and then the rate of combustion increases. This is accompanied by collapse in the core and an expansion of the hydrogen-burning surface layers, forming a RED GIANT. Progressive gravitational collapses and burning of the helium (generated by the consumption of the hydrogen) result in a WHITE DWARF which is a sphere of enormously dense gas. The white dwarf cools over many millions of years and forms a black dwarf—an invisible ball of gases in space. Other sequences of events may occur depending upon the size of the star formed. BLACK HOLES and NEUTRON STARS may form from red (super) giants via a SUPERNOVA stage.

stellar wind a similar phenomenon to the SOLAR WIND—an outpouring of material from a hot star.

stereochemistry the part of chemistry that covers the arrangement of atoms in space within a molecule (see ISOMER).

stereotaxis the movement or reaction of an organism due to its contact with a solid body.

steroids a group of LIPIDS (fats) with a characteristic structure comprising four carbon rings fused together. The group includes the sterols (e.g. CHOLESTEROL), the BILE acids, some HORMONES, and vitamin D. Synthetic steroids act like steroid hormones and include derivatives of the glucocorticoids used as anti-inflammatory agents in the treatment of rheumatoid arthritis; oral contraceptives which are commonly mixtures of OESTROGEN and a derivative of PROGESTERONE (both female sex hormones); anabolic steroids e.g. TESTOSTERONE, the male sex hormone, which is used to treat medical conditions such as osteoporosis and wasting. However, much publicity surrounds the use of the anabolic steroids by athletes, contrary to the rules of sports-governing bodies, to increase muscle bulk and body weight.

stoichiometry an aspect of chemistry that deals with the proportion of elements (or chemical equivalents) making pure compounds.

stony meteorites meteorites composed mainly of rock-forming SILICATES including PYROXENE, OLIVINE and plagioclase (feldspar) with some nickel-iron. This type accounts for the vast majority of meteorites that are *seen* to fall. They are termed either chondrites or achondrites depending upon the presence, or lack, of chondrules which are glassy droplets, up to 2 mm in size, produced by the melting and sudden cooling of silicate material.

strain when forces acting upon a material produce distortion, it is said to be strained, or in a state of strain. Strain is represented as a ratio of the change in dimension or volume to the original dimension or volume.

strain gauge a device to measure strain. A basic version comprises a fine metal or semiconductor grid on a backing sheet which is fastened to a body that is to be subjected to strain. The grid is then strained by the same amount, altering the electrical properties of the grid, which can be measured. There are many other strain gauges which employ light, vibrations/sound and the use of liquid crystals.

stratigraphy is a branch of geology that involves the study of stratified, or layered rocks with particular regard to their position in time and space, their classification and correlation. This may entail the use of fossils, characteristic lithologies and/or time intervals.

stratocumulus a grey or white CLOUD composed of sheets or layers usually with dark patches.

stratopause the top of the stratosphere at about 50 kilometres.

stratosphere the layer of the atmosphere above the TROPOSPHERE which stretches from 10 to 50 km above the ground. It is a stable layer with the TROPOPAUSE at the base. The temperature increases from the lower part to the upper, where it is 0°C and the higher temperatures occur as a result of ozone absorbing ultraviolet radiation. The inversion of temperatures thus creates the stability which tends to limit the vertical extent of cloud, producing the lateral extension of, for example, CUMULONIMBUS cloud into the anvil head shape.

stratum (plural *strata*) in geology a layer or bed of rock which has no limit on its thickness.

stratus a spread out cloud form with an even base and generally grey appearance through which the Sun may be seen, providing the cloud is not too dense. It occasionally forms ragged patches.

stress is force per unit area. When applied to a material, a corresponding STRAIN is created. The two main types are tensile (or compressive) and shear stress. Units, typically, include: kNm^{-2} (kilonewtons per square metre), $lbfin^{-2}$ (pounds force per square inch).

striation (or striae) a product of glaciation which results in lines or grooves being scratched on the surface of exposed rock due to the action of hard rocks embedded in the base of the glacier. The lines may provide a guide to the direction of ice movement in areas which have undergone glaciation.

strike the direction of a horizontal line, measured with a compass, on an inclined plane e.g. a rock bedding plane. It is perpendicular to the dip. Measurement of strike and dip is a vital procedure for geologists undertaking mapping, as it allows them to plot the results and determine large scale trends and structures.

stroboscope an instrument used to view rapidly moving objects by shining a flashing light source onto an object that is revolving or vibrating, the object can be made to appear still, providing the frequency of the light flashes match the revolutions or vibrations. This is a technique used in engineering to examine engine components.

stroma (*plural* **stromata**) any tissue that functions as a framework in plant cells (*see* CALVIN CYCLE).

strophism the twisting of a stalk as it grows in response to a stimulus from a particular direction e.g. light.

structural formula a formula providing information on the ATOMS present in a MOLECULE and the way that they are bound together, i.e. an indication of the structure.

subduction zone an essential component of the concept of PLATE TECTONICS. It is the area where a lithospheric plate descends beneath continental crust at deep trenches in the ocean, thus returning old LITHOSPHERE material to the MANTLE. It is therefore known as a destructive plate boundary. Evidence for the process comes from evidence of magnetic fields preserved in the rocks of the ocean floor. A zone of earthquake foci is associated with the subducted slab. Volcanic activity is also a feature found near subduction zones.

sublimation is when a vapour forms directly from a solid, without going through the liquid phase.

substitution (reaction) a reaction in which an atom or group in a molecule is replaced by another atom or group, often hydrogen by a HALOGEN, hydroxyl, etc.

substrate in biology the substrate is the surface upon which an organism lives and from which it may derive its food. It is also a substance in a reaction which is catalysed by an ENZYME. In electronics, it is the single crystal or SEMICONDUCTOR used as the base upon which an integrated circuit or TRANSISTOR is printed.

sucrose a disaccharide CARBOHYDRATE ($C_{12}H_{22}O_{11}$) occurring in beet, sugar cane and other plants (see SACCHARIDES). It consists of the two monosaccharides fructose and glucose and is the form in which plants carry carbohydrates from one part of the plant to another.

sugar a crystalline monosaccharide or oligosaccharide (a small number, usually two to ten, monosaccharides linked together, with the loss of water), soluble in water. Also, it is used generally as the common name for SUCROSE.

sulphur a yellow, non-metallic element which exhibits allotropy (that is an element which exists in several forms with different physical properties). It is widely distributed both in the free state and in compounds, especially as sulphates (e.g., GYPSUM) and sulphides (e.g. PYRITE) and as hydrogen sulphide (H_2S) in natural gas and oil. It is manufactured by heating pyrite or purifying the naturally-occurring material. Its primary use is in the manufacture of SULPHURIC ACID but it is also used in the preparation of matches, fireworks, dyes, fertilizers, fungicides etc.

sulphuric acid (H_2SO_4) a strong acid that is highly corrosive and reacts violently with water, with the generation of heat. It is manufactured by a process called Contact process, from sulphur or ores of sulphur. It is used widely in industry in the manufacture of dyestuffs, explosives, other acids, fertilizers, and many other products.

Sun the star nearest to Earth and around which the Earth and the other planets rotate in elliptical orbits. The Sun has a diameter of 1.392×10^6 kilometres and its mass is approximately 2×10^{30} kilograms. The interior reaches a temperature of 13 million degrees Centigrade, while the visible surface is about 6000°C. The internal temperature is such that thermonuclear reactions occur, converting HYDROGEN to HELIUM with the release of vast quantities of energy. The Sun is approximately 90 per cent hydrogen, 8 per cent helium, and is 5 million years old—roughly halfway through its anticipated life cycle.

sunshine recorder (Campbell-Stokes recorder) an apparatus comprising a glass sphere which focuses the Sun on card marked with hours. The focusing creates sufficient heat to burn a track on the card, thus recording the duration of the sunshine.

sunspots the appearance of dark areas on the surface of the SUN. The occurrence reaches a maximum approximately every eleven years in a phase known as the sunspot cycle. They are usually short-lived (less than one month) and are caused by magnetism drawing away heat to leave a cooler area which is the sunspot. The black appearance is due to a lowering of the temperature to about 4000K. Sunspots have intense magnetic fields and are associated with magnetic storms and effects such as the *aurora borealis* (see AURORA). They may send out solar flares which are explosions occurring in the vicinity of the sunspots.

superconductivity the property of some metals and alloys whereby their electrical RESISTANCE becomes very small around ABSOLUTE ZERO. The potential uses for superconductors include circuits in large computers, where superconductive circuits would generate far less heat than is the case presently. This would enable larger and faster computers to be built. Transmission of electricity and reduction of the associated heat loss is also under study.

superfluidity is when a fluid flows without friction. This is shown by helium-4 (^{4}He) below 2.19K when it behaves with zero effective viscosity. The temperature is called the lambda point and at temperatures above, ^{4}He is termed helium I and helium II below.

supernova (*plural* **supernovae**) a star which explodes, it is thought, due to the exhaustion of its hydrogen (see SUN), whereupon it collapses, generating high temperatures and triggering thermonuclear reactions. A large part of its matter is flung into space, leaving a residue that is termed a WHITE DWARF star. Such events are very rare, but at the time of explosion, the stars become one hundred million times brighter than the Sun. A supernova was seen in the large Magellanic Cloud in 1987, even though it is 170 000 light years away.

supersaturation the state of a SOLUTION when it contains more dissolved SOLUTE than is required to produce a SATURATED SOLUTION.

surd the root of quantity that can never be exactly expressed because it is an IRRATIONAL NUMBER, e.g. $\sqrt{3} = 1.73205$.

surface tension is the 'tension' created by forces of attraction between molecules in a liquid, resulting in an apparent elastic membrane over the surface of the liquid. This attraction between molecules of the same substance is called *cohesion* and the result is that it tries to pull liquids into the smallest possible shapes. This can plainly be seen in water which forms round droplets and also supports the feet of insects on ponds and puddles. The same phenomenon is demonstrated by a needle on a piece of blotting paper which is then placed gently on water. When the paper absorbs sufficient water to sink, the needle remains afloat, because of the surface tension of the water. Droplets of mercury show the same effect, forming compact globules on a surface.

surface wind the wind as measured close to the Earth's surface, actually at a height of 10 metres. The speed of the surface wind is reduced by friction of the surface.

surfactant (*also called* **surface-active agent**) a compound that reduces the SURFACE TENSION of its SOLVENT, e.g. a detergent in water.

suspension a two-phase system with particles distributed in a less dense liquid or gas. Settling is prevented/slowed by the VISCOSITY of the fluid or impacts between the molecules of the fluid and the particles themselves.

symbiosis is a relationship between organisms, usually from two different species, which has beneficial consequences for at least one of the organisms. There are various forms of symbiosis, including commensalism, where one party benefits but the other remains unharmed, and parasitism, where one party greatly benefits (the PARASITE) at the other party's expense. Symbiosis can also solely refer to mutualism, where both parties benefit and neither is harmed. There are many examples of parasitism, including nematodes, worms in humans. Certain birds form the active party in commensalism where, for example, they feed on insects on or around a larger animal. The digestion of cellulose by microorganisms in the gut of ruminant animals is an example of mutualism.

symbols in chemistry are used to represent elements, atoms, molecules, etc. Each element has its own symbol of one or two letters (see PERIODIC TABLE), thus fluorine is F, and chlorine is Cl. Symbols are used further in formulae, i.e. the shorthand representation of a compound, for example NaCl is sodium chloride. The formulae can then be used in equations to represent chemical reactions and processes, e.g.

$$NaOH + HCl \rightarrow NaCl + H_2O$$

This formula states that caustic soda and hydrochloric acid, when combined, will react to form sodium chloride and water.

Symbols are used elsewhere, particularly in the sciences, to provide a convenient shorthand, e.g. prefixes in decimal numbers (see SI UNITS), concepts in physics such as m for mass and I for electric current and in mathematics where a letter or figure represents a word or sentence.

symmetry is the property of a geometrical figure whose points have corresponding points reflected in a given line (axis of symmetry), point (centre of symmetry) or plane. Symmetry is closely related to balance in nature and many forms exhibit bilateral symmetry, humans included. Symmetry is very evident in crystals that have grown in ideal conditions because then the crystals faces are apparent and most crystals exhibit several symmetrical features (see *also* CRYSTALLOGRAPHY).

synapse (see *also* NEURON) the junction between two nerve cells where a minute gap (of the order of 15 NANOMETRES) occurs. The nerve impulse is carried across the gap by a chemical called a neurotransmitter. The chemical diffuses across the gap connecting the axon of one nerve cell to the dendrites of the next. An individual neuron commonly has several thousand such junctions with around 1000 other neurons.

synergism is when the combined effect of two substances (e.g. drugs) is greater than expected from their individual actions added together.

synthesis the formation of a compound from its constituent elements or simple compounds.

systolic blood pressure the pressure generated by the left VENTRICLE of the HEART at the peak of its contraction. Since the left ventricle has to pump blood to all parts of the body, it generates a higher pressure than the right ventricle, which pumps blood only to the lungs. In normal people, the systolic blood pressure is 120 mm of mercury (120 mm Hg), and when the ventricle relaxes, pressure is still maintained in the blood vessels. This resting pressure is called the diastolic pressure and is approximately 80 mm Hg. This is the familiar blood pressure measurement and is represented as 120/80. This fluctuation in pressure is responsible for the pulse, which also represents the heartbeat.

T

tangent a function of an angle in a right-angled triangle, defined as the ratio of the side opposite the angle to the length of the side adjacent to it. Also, in geometry, a straight line that just touches the circumference of a circle.

tartaric acid an organic acid which exists in four forms that are STEREOISOMERS. The commonest form (d-tartaric acid) is used in dyeing, the manufacture of baking powder and "health salts." Another form (dl-tartaric acid, or racemic acid) occurs in grapes.

tautomerism a special case of structural isomerism (see ISOMER), often called dynamic isomerism. It is when a compound exists as a mixture of two forms, or isomers (in equilibrium). The two forms can each change to the other as when, for example, one isomer is extracted from the mixture, then some of the other isomer will change to re-establish equilibrium. The ability to make this reversible change is due to a mobile atom or group, often hydrogen, which alters its position in the molecule possibly with the changing of a double bond. Isomers produced in this way are called TAUTOMERS and each can give rise to stable derivatives. A typical example would be:

$$> C - C = C \Longleftrightarrow > C = C - C <$$

taxis (*plural* **taxes**) the movement of a cell or organism in response to a stimulus in environment. This stimulus may be temperature (thermotaxis), light (phototaxis), gravity (geotaxis) or chemical (CHEMOTAXIS).

taxonomy the study, identification and organization of organisms according to their similarities and differences. Taxonomy is concerned with the classification of all organisms, whether plant or animal, dead or alive and so fossils are also important. Modern taxonomy provides a convenient method of identification and classification of organisms, which expresses the evolutionary relationships to one another. There are different levels of classification with five kingdoms, beneath which are the phyla, classes, right down to the level of species.

T-cell a type of white blood cell (LYMPHOCYTE) that forms sin the bone marrow and moves to the thymus gland, situated in the thorax. There are a whole variety of T-cells involved in recognizing particular foreign bodies (ANTIGENS), and they are important in fighting viral infections and destroying bacteria that have penetrated the cells of the body.

tectonic concerned with earth movements, as involved in FOLDing and FAULTing.

telephone is an instrument that enables speech to be transmitted by means of electric currents or radio waves. It was invented by Alexander Graham Bell in 1876 and a public service was begun three years later after Bell brought his invention to the UK. The modern telephone consists of a mouthpiece containing a thin diaphragm of aluminium which moves with the sound of speech. This movement presses carbon granules which produces a surge of current and in the earpiece of the receiving set, these surges are changed back into sound. An electromagnet reacts to the current charges and vibrates a diaphragm, thus reproducing the voice of the speaker.

Advances in technology have improved the transmission of telephone calls, satellites are used for international calls and recently, optical fibres have been introduced as a new medium of transmission.

telescope is an instrument for magnifying an image of a distant object, the main types of astronomical telescopes being *refractors* and *reflectors*. The refracting type have lenses to produce an enlarged, upside-down image. In the reflecting type there are large mirrors with a curved profile which collect the light and direct it onto a second mirror and into the eyepiece.

telophase the last and fourth stage of MEIOSIS or MITOSIS in EUCARYOTIC cells. During telophase, a nuclear membrane forms round each of the two sets of CHROMOSOMES that have formed separate groups at the spindle poles. The chromosomes decondense, the nucleoli reappear, and the cell eventually splits to form two daughter cells.

temperature degree of heat or cold against a standard scale.

terminal velocity the constant VELOCITY achieved by a body falling through a medium when the pull of GRAVITY is equalled by the frictional resistance.

terpenes colourless, liquid hydrocarbons that occur in many fragrant natural oils of plants. The general formula is $(C_5H_8)n$ where C_5H_8 is the basic isoprene unit. This leads to their classification: monoterpenes are $C_{10}H_{16}$; sesquiterpenes $C_{15}H_{24}$; diterpenes $C_{20}H_{32}$, and so on.

terpenoids a group of SECONDARY METABOLITES found in plants and based on isoprene units (C_5; see *also* TERPENES). The group includes many essential oils, the CAROTENOIDS, rubber, and the gibberellins which are plant growth substances, promoting shoot elongation in some plants and promoting seed germination.

testosterone a male sex hormone that promotes the development of male characteristics. It belongs to the group of sex hormones called androgens, which are produced mainly by the testes. Testosterone is responsible for triggering growth of facial and pubic hair and development of the testes.

thermal conductivity a measure of the rate of heat flow along a body by conduction.

thermal metamorphism (or contact metamorphism) the process of metamorphism which occurs when rocks are intruded by an igneous body and the rocks are recrystallized in response to the heat. There is little or no associated pressure, thus large scale structures may remain while the mineralogy changes. The zone of altered rock is called the metamorphic AUREOLE, and clearly the greatest changes are observed next to the intrusion. PELITIC rock assemblages show the greatest sensitivity to the temperature and new minerals will form e.g. cordierite (a silicate with aluminium, iron and magnesium) and andalusite (aluminium silicate, Al_2SiO_5).

thermistor a temperature-sensitive SEMICONDUCTOR whose RESISTANCE decreases with an increase in temperature. Thermistors are used for temperature measurement and compensation.

thermochemistry the branch of chemistry dealing with the heat changes of chemical reactions.

thermocouple a device which is used for measuring temperature, which consists of two wires made of different metals, joined at each end. The temperature is measured at one join, and the other join is kept at a fixed temperature. A temperature difference between the two joins creates a voltage which causes a current to flow. Either the voltage or the current can be measured, thus creating a calibrated device. The effect is due to the movement of electrons because of the temperature difference. The two metals used are often copper and constantan (a copper/nickel alloy) which has a constant resistance, whatever the temperature.

thermodynamics is the study of laws affecting processes that involve heat changes and energy transfer. Heat transfer from one body to another, the link between heat and work and changes of state in a fluid all come within the field of thermodynamics, it is the prerequisite to analysis of work by machinery. There are essentially three *laws of thermodynamics*. The first law says that heat is a form of energy and is conserved and any work energy produced in a closed system must arise from the conversion of existing energy, i.e. energy cannot be created or destroyed. The second law states that the entropy of any closed system cannot decrease and if the system undergoes a reversible process it remains constant, otherwise it increases. The result of this is that heat always flows from a hot body to a cooler one. The third law states that absolute zero (see TEMPERATURE) can never be attained.

thermograph a thermometer that records continuously, and where the recording is usually achieved by means of an electrical device.

thermography the medical scanning technique in which the INFRARED RADIATION or radiant heat emitted by the skin is photographed, using special film, to then create images. An increase in heat emission represents an increase in blood supply, which may be a sign of a CANCER. The technique is used to detect cancers, especially of the breast.

thermoluminescence a phenomenon whereby a material gives out light upon heating, due to ELECTRONS being freed from DEFECTS in crystals. The defects are generally due to ionizing radiations, and the principle is applied in dating archaeological remains, especially ceramics, on the assumption that the number of trapped electrons, caused by exposure to radiations, depends upon time. Although this is not absolutely correct, an estimate of the age of a piece of pottery can be obtained by heating and comparing the thermoluminescence with that of an item of known age.

thermolysis breakdown of a compound or molecule by heat.

thermometer an instrument used to measure temperature. The basis of a thermometer is a property of a substance that varies reliably with temperature, e.g. expansion. Thermometers that utilise a liquid in glass are based upon the property that liquids expand slightly when they are heated. Both mercury and alcohol are used and when the bulb at the base of the thermometer's stem is heated, the liquid expands up the stem to create a reading. More sensitivity is gained by using a narrower tube. This is the case with *clinical thermometers* where the scale covers just a few degrees on either side of the normal body temperature of 37°C.

In industry, thermocouples are used for temperature measurement in furnaces, and other instruments that provide an electrical measurement are used in preference to liquid in glass thermometers which have a limited range. *Resistance thermometers* are based on the property that the electrical resistance of a conductor normally increases with heat and so it becomes more difficult to pass an electric current. A spiral of platinum wires is used in this case. A *thermistor thermometer* works on the same principle, but consists of a semiconductor in which the resistance decreases with a temperature increase, e.g. 100,000 ohms at 20°C and just 10 ohms at 100°C.

thermoplastic a plastic material that can be melted or softened by heat and then cooled, repeatedly, without significant alteration in its properties. Thermoplastics include polystyrenes and vinyl polymers.

thermostat a device used for keeping a constant temperature through the supply or non-supply of heat when the required temperature is not achieved. Thermostats are used in ovens, immersion heaters, refrigerators and a common type consists of a bimetal strip. This has two thin strips of different metals fastened together and on heating there is a difference in expansion, making the strip bend. This can then be used to operate a switch.

thixotropy the property of some fluids that are very viscous (syrup-like) until a STRESS is applied, when the fluid flows more easily. This principle is commonly utilized in the manufacture of non-drip paints.

thrust a fault inclined at a low angle (usually less than 45°) in which the sense of movement is reverse, that is one block moves up the thrust plane over the underlying block.

thunder the rumbling noise that accompanies lightning flashes which is due to the sudden heating and expansion of the air caused by the electrical discharge. The continuing noise is due to sound travelling from the various sections of the discharge - because the spark can be several kilometres long.

thyroid gland *see* **endocrine system**.

tide tides affect the surface layers of a planet (or natural satellite) whether liquid or solid, due to the effects of gravitational forces. The ocean tide on Earth is due to the attraction of mainly the Moon, but also of the Sun and the resulting effect is a regular rise and fall of water levels in the oceans and seas. Variation in tides is caused by the positions of the three bodies and when the Moon and the Sun "pull" in the same direction, there is a high spring tide; when they are at 90°, there is a low neap tide. Other factors which affect tides are the uneven distribution of water on the Earth's surface and the topography of the sea bed. The effect of tides in the open oceans is not very noticeable but in shallow seas tides may be several metres.

till sediment deposited due to the action of glacial ice without water as an agent. The size of the particles varies from clay particles to rock fragments. There is a variety of tills depending upon their method of release, for example, subglacial melting gives lodgement till and the thawing of stationary ice produces melt-out till.

time zones zones which run north-south, with some variations across the Earth, and which represent different times. Each zone is one hour earlier or later than the adjacent zone, and is 15° of *longitude*. The zones were devised for convenience, but to compensate for the accumulated time change, the *International Date-Line* was introduced. The line runs roughly on the 180° meridian, although it does detour around land areas in the Pacific Ocean, and to cross it going east means repeating a day, while in the opposite direction losing a day.

tin a soft, malleable and ductile metal (SYMBOL Sn) which exhibits allotropy (occurring in different physical forms). It occurs naturally as oxides and is used to coat steel and in producing alloys (solders, fusible alloys, etc.). It is obtained easily from its ores (e.g., cassiterite, SnO_2) and was known thousands of years ago.

tissue a group of cells with a similar function which aggregate to form an organ.

tissue culture the culture or growth, outside the body, of TISSUES of living organisms. The cells are placed in an artificial medium containing nutrients and other factors such as temperature and pH are controlled, while waste products are removed. The technique has proven valuable in studying cell growth and it has been applied to the propagation of plants.

titanium (Ti) a malleable and ductile metal that resembles iron. The main source is the ore rutile (TiO_2). It is light, strong and highly resistant to corrosion. It is therefore useful in aircraft and missile manufacture. It is quite widely distributed in rocks and is also found in meteorites. Its other uses include catalysis and as a white pigment in the form of the oxide TiO_2.

titration the laboratory procedure of adding measured amounts of a SOLUTION to a known volume of a second solution until the chemical reaction between them is complete. If the concentration of one solution is known, and both volumes are known then the unknown strength of one solution can be determined. The end of the chemical reaction between the solutions is found by use of an indicator which changes colour at the appropriate point.

tomography a scanning technique that uses X-rays for photographing particular "slices" of the body. A special scanning machine moves around the patient who is lying flat, taking measurements every few degrees over 180°. The scanner's own computer builds up a three-dimensional image that can then be used for diagnosis. Such a technique has the dual benefit of providing more detail than a conventional X-ray and yet delivers only one fifth of the dose.

tonne a metric ton (1000kg).

tornado a narrow column of air that rotates rapidly and leaves total devastation in its path. It develops around a centre of very low pressure with high velocity winds (well over 300 km/hour) blowing anticlockwise and with a violent downdraught. The typical appearance is of a funnel or snake-like column filled with cloud and usually no more than 150 metres across.

Tornadoes are often unpredictable in their behaviour and can lose contact with the ground or retrace its route. When it moves out over the sea, and once the tunnel has joined with the waves, a *waterspout* is formed.

torus a 'doughnut' shaped ring, which is generated by rotating a circle about an axis.

trace element an ELEMENT that occurs in very small quantities in rocks, but which can be detected by geochemical analysis. All elements but the most commonly occurring form trace elements and in quantities much less than 1%. Very often trace elements occur in minute quantities ranging from a few parts per million to several hundred. In biology, trace elements are those required by an organism in very small quantities but nevertheless they are necessary for good health.

trace fossil a SEDIMENTARY STRUCTURE due in some way to the presence or activity of an organism. The study of trace fossils is called ichnology. They are most common at the junction of different lithologies (see LITHOLOGY), for example shale and sandstone. The fossils may occur as ridges, tubes, burrows etc.

trade winds these play an important part in the atmospheric circulation of the Earth and they are mainly easterly winds that blow from the subtropics to the equator. The *Westerlies* flow from the high pressure of the subtropics to the low pressure of the temperate zone. The Westerlies form one of the strongest wind flows and their strength increases with height (see JET STREAM) and depressions are most common in this wind system. The *Doldrums* is a zone of calms or light winds around the equator applied particularly to the oceans, with obvious links to the time when sailing ships were becalmed. Also linked to sailing are the *Roaring Forties*, which are Westerlies in the southern hemisphere where they tend to be stronger. However, the supposed link of trade wind with early travel on the sea is incorrect— their origin is from the latin word meaning constant.

transcription the formation of an RNA molecule from one strand of a DNA molecule. Transcription involves many processes, starting with the unwinding of the double-stranded DNA helix, along which an enzyme, called RNA polymerase, travels and catalyses (see CATALYST) the formation of the RNA molecule by pairing NUCLEOTIDES with the corresponding sequence of the DNA strand. As the RNA molecule leaves, the DNA reforms its double-stranded helix.

transducer a device that converts one form of energy into another, often a physical quantity into an electrical signal, as in microphones and photocells. The reverse also applies, as in loudspeakers.

transferase an ENZYME which catalyses the transfer of chemical groups between molecules e.g. acyl transferase.

transfer RNA (tRNA) one of the three major classes of RNA that functions as the carrier of AMINO ACIDS to RIBOSOMES, where the POLYPEPTIDE chains of PROTEINS are formed. Every tRNA molecule has a structure that will accept only the specific attachment of one amino acid.

transformer a device for changing the VOLTAGE of an ALTERNATING CURRENT. The unit consists essentially of an iron core with two coils of wire. Current fed into the primary coil generates a current in the secondary through ELECTROMAGNETIC INDUCTION. The ratio of the voltage between the coils is determined by the ratio of the number of turns in each coil. Transformers can therefore be used to step down or step up a voltage and they play a vital role in the transmission of mains power across the country. Huge transformers are used in the mains power supply between the power station and the domestic supply. Current generated at a power station goes through a step-up transformer, creating voltages of up to 400,000 volts, at much reduced currents (thus minimizing heat loss) for transmission through the power lines of the *grid*. Power from the grid then goes to substations where transformers step the voltage down by a series of transformers to 132,000 volts, then 33,000v (for heavy industry), then 11,000v for light industry and finally 240v for offices and homes.

transform fault a basic component of the theory of PLATE TECTONICS because transform faults permit the subdivision of the Earth into plates undergoing relatively little internal deformation. Numerous transform faults cut across mid-oceanic ridges.

transgression the result of an increase in sea level producing an advance of the sea over new land areas. Thus deeper water sediments are deposited over shallow water sediments.

transit movement of a small body across the disk of a larger body (to an observer on Earth) as with a satellite moving across its parent planet.

transition point the point at which a substance may exist in more than one solid form, in equilibrium.

transition element one of the ELEMENTS in the periodic table that is characterized by an incomplete inner electron shell and a variable valency. They are metallic in nature, the chemical properties of one element resemble those of the adjacent element in the PERIODIC TABLE.

transistor a SEMICONDUCTOR device that is used in three main ways: as a switch; a rectifier (or DIODE, which conducts current in one direction, thus turning AC into DC); and as an amplifier creating strong signals from weak ones. It consists basically of a semi-conductor chip of silicon which is doped to form two p-n junction diodes, back to back. Current only flows through when a small current is applied to the p-type region of the semi-conductor called the base circuit. When this current is applied, an enlarged current flows in the output or collector circuit and this property allows it to be used as a switch, because if there is no current in the base circuit, a current cannot flow in the collector circuit.

translation the synthesis of PROTEINS in a RIBOSOME that has MESSENGER RNA (mRNA) attached to it. As the mRNA molecule moves through part of the ribosome, a TRANSFER RNA molecule carrying the appropriate AMINO ACID will enter a site on the ribosome and will be released after it has contributed a new amino acid to the growing chain.

transpiration the loss of water vapour from pores (stomata) in the leaves of plants. Transpiration can sometimes account for the loss of over one sixth of the water that has been taken up by the plant roots. The transpiration rate is affected by many environmental factors—temperature, light and carbon dioxide (CO_2) levels, air currents, humidity and the water supply from the plant roots. The greatest transpiration rate will occur if a plant is photosynthesizing (see PHOTOSYNTHESIS) in warm, dry and windy conditions.

transpose in mathematics, to move one term or element from one side of an equation to the other with a corresponding reversal in sign. A MATRIX formed from another by interchanging the rows and columns: the transpose of matrix A is usually denoted AT.

transverse wave a wave in which the vibration occurs at right angles to the direction of wave propagation, e.g. an ELECTROMAGNETIC WAVE or, more simply, a wave on a taut piece of string.

trapezium a QUADRILATERAL (a four-sided plane figure) with two parallel sides (see *also* ISOSCELES).

tribology the study of FRICTION, lubrication and wear, as occurs when two surfaces are in contact in relative motion. It includes the study of substances that reduce wear, overheating, etc. in such circumstances, and is important in the efficient operation of engines and machines.

trigonometric function one of the functions, such as $\sin(x)$, $\tan(x)$ and $\cos(x)$, obtained from studying certain ratios of the sides of a right-angled triangle.

trigonometry the study of right-angled triangles and their TRIGONOMETRIC FUNCTIONS.

trisomy the abnormal condition in which an organism has three CHROMOSOMES rather than the normal pair for one type of chromosome. Trisomy can occur in humans and results in offspring with abnormal characteristics and shorter-than-average lifespans. One common example of trisomy is Down's syndrome, caused by the presence of three instead of two chromosomes of the number 21 type (all the other chromosomes are in normal pairs).

trophic a term meaning relating to nutrition

tropism growth of a plant organ in a particular direction due to an external stimulus e.g. touch, light.

tropopause the top of the TROPOSPHERE. It is the boundary between the troposphere and the stratosphere. The height of the tropopause differs over short periods, from 9-12 km over the poles to about 17 km over the equator.

troposphere is the Earth's atmosphere between the surface and the TROPOPAUSE. This layer contains most of the water vapour in the atmosphere and most of the AEROSOLS in SUSPENSION. The temperature decreases with height at approximately 6.5°C per kilometre and it is in this layer that most weather features occur.

tsunami the giant sea waves produced by sudden large scale movements of the sea floor whether due to EARTHQUAKE, volcanic explosion or enormous slides and slumps. It is likely that the slides and slumps are themselves triggered by an earthquake. The effect of the sea floor displacement is not especially apparent in the open oceans but the waves, which can travel at speeds of several hundred kilometres per hour, contain an immense amount of energy. Consequently, when the force of the water reaches the shallows or narrow inlets, the effects are catastrophic. A susceptible area is the western coast of the Pacific (Japan, etc.) where there are deep water trenches associated with zones of earthquakes and highly populated islands.

tundra is the treeless region between the snow and ice of the Arctic and the northern extent of tree growth. Large treeless plains can be found in northern Canada, Alaska, northern Siberia and northern Scandinavia. The ground is subject to PERMAFROST but the surface layer melts in the summer, so soil conditions are very poor, being waterlogged and marshy. The surface therefore can support little plant life. Cold temperatures and high winds also limit the diversity of plants, restricting the *flora* to grasses, mosses, lichens, sedges and dwarf shrubs. Some areas of tundra receive the same low level of precipitation as deserts yet the soil remains saturated due to the partial thaw of the permafrost. Most growth occurs in rapid bursts during the almost continuous daylight of the very short summers.

In addition to this Arctic tundra, there is also *alpine tundra* which is found
on the highest mountain tops and is therefore widely spread. However, conditions differ because of daylight throughout the year and plant growth in the
tropical alpine tundra also occurs all year round.

tungsten a hard grey metal used in alloys of steel where its hardness and resistance to corrosion are useful. It is used in CARBIDE tools and electric lamp filaments. Tungsten carbide is almost as hard as diamond and is used extensively in
abrasives.

twilight the period of partial light after sunset or before sunrise caused by the
reflected sunlight in the upper atmosphere when the Sun is below the horizon.
In astronomy, it is defined as beginning when the Sun is 18° below the horizon.

typhoon *see* **hurricane**.

U

ultrabasic rock an IGNEOUS ROCK that contains many silicate minerals of iron and magnesium such as AMPHIBOLES and PYROXENES, and with SILICA occurring as quartz.

ultracentrifuge a special centrifuge machine that generates high centrifugal forces as it is capable of spinning at speeds of up to 50,000 revolutions per minute. The ultracentrifuge is most commonly used during the separation of the various ORGANELLES within cells. The larger and more dense organelles will form a deposit in the centrifuge tube more readily than the smaller, less dense ones. The nucleus is the largest organelle of any normal cell and will be deposited at the bottom of a centrifuge tube when the ultracentrifuge spins at a force of 600g for 10 minutes, whereas the smaller MITOCHONDRION needs a higher speed of 15,000g for 5 minutes to be deposited.

ultrasonic a term used to describe sound waves that are inaudible to humans as they have a frequency above 20kHz. Although the human ear is incapable of detecting such a high FREQUENCY, some animals, such as dogs and bats, can detect ultrasonic waves (*also known as* **ultrasound**). Ultrasound is used widely in industry, medicine and research. For example, it is used to detect faults or cracks in underground pipes and to destroy kidney stones and gallstones. The most recent development in ultrasonics is their use in chemical processes to trigger reactions involved in the production of food, plastics and antibiotics. Ultrasonics make certain chemical processes safer and cheaper as they eliminate the need for high temperatures and expensive catalysts.

ultraviolet radiation a form of radiation that occurs beyond the violet end of the visible light spectrum of ELECTROMAGNETIC WAVES. Ultraviolet rays have a FREQUENCY ranging from 10^{15}Hz to 10^{18}Hz, with a wavelength from 10^{-7}m to 10^{-10}m. They are part of natural sunlight and are also emitted by white-hot objects (as opposed to red-hot objects, which emit INFRARED RADIATION). As well as affecting photographic film and causing certain minerals to fluoresce, ultraviolet radiation will rapidly destroy bacteria. Although ultraviolet rays in sunlight will convert steroids in human skin to vitamin D (essential for healthy bone growth), too much can cause irreversible damage to the skin and eyes and damage the structure of the DNA in cells. Fortunately, a great deal of the ultraviolet radiation from the sun is prevented from reaching the earth as the OZONE LAYER in the upper atmosphere acts as a UV filter.

umbra a region of complete shadow, which is usually applied to eclipses. The area of half or partial shadow is termed the penumbra.

unconformity a break in the deposition of sedimentary rocks, allowing erosion of previously formed rock before eventual deposition of further sediments. It is usually represented by an obvious difference in the attitude of the rocks on either side of the unconformity, with the upper lying unconformably on the lower.

unified scale (see also RELATIVE ATOMIC MASS) the scale which lists atomic and molecular weights using the ^{12}C isotope as the basis for the scale and taking the mass of the isotope as exactly twelve. This means the atomic mass unit is 1.660 x 10^{-27}kg.

unit cell the smallest fragment of a CRYSTAL that will reproduce the original crystal if the unit cells are arranged in a repeating, three dimensional pattern.

unit vector a VECTOR with a magnitude of one.

unity the number or numeral one; a quantity assuming the value of one.

universe all matter, energy and space in the cosmos. The size of the universe that can be observed is limited by the distance light has travelled since the Big Bang.

universal gas equation see **gas laws**.

universal indicator a mixture of certain substances, which will change colour to show the changing pH of a SOLUTION. Universal indicator is available in the form of a solution or paper strip and is used as an approximate measure of the pH of a solution by using the following chart as a guide:

Colour of indicator	red	orange	yellow	green	blue	purple
pH	1 2 3	4 5 6	7 8 9	10 11	12 13	14

unsaturated a chemical term that can be applied to a compound or solution. In the case of unsaturated organic compounds, the carbon atoms are called unsaturated as they form double or triple bonds and are thus capable of undergoing reactions which can add groups or atoms and break the double or triple bonds to form single bonds. If a SOLUTION is described as unsaturated, then it contains a lower concentration of SOLUTE dissolved in a definite amount of solvent than would be found in a SATURATED solution.

uranium a metallic element that is radioactive and has the greatest mass of all naturally occurring elements (atomic mass of 238). Uranium has 92 protons within its nucleus and exists as three ISOTOPES, ^{238}U, ^{235}U, and ^{234}U—each of which undergoes ALPHA DECAY. Uranium will naturally disintegrate and pass through a series of other elements to form eventually a stable isotope of the element lead. When uranium is bombarded with NEUTRONS, however, it undergoes artificial disintegration to form two other heavy nuclei, releasing a very large amount of energy—this is the basic process underlying nuclear FISSION, which is used to generate energy in nuclear power stations. Uranium is not only used as a fuel in nuclear reactors but is also used in atomic bombs. Indeed, the first atomic bomb, dropped on Hiroshima in 1945, is believed to have contained two or more small quantities of the isotope ^{235}U, which were suddenly brought together by a device and the CHAIN REACTION of nuclear fission immediately ensued.

Uranus is the seventh planet in the SOLAR SYSTEM and one of the four *gas giants* with an orbit between those of Saturn and Neptune. Uranus has a diameter at the equator of 50,080 kilometres and lies an average distance of 2,869,600,000 kilometres from the Sun. The surface temperatures are in the region of -240°C. It is composed mainly of gases with a thick atmosphere of methane, helium and hydrogen. Uranus was the first planet to be observed with a telescope and was discovered by William Herschel, a German astronomer in 1781. Uranus remained a mystery until quite recently in 1986, *Voyager 2* approached close to the planet and obtained valuable information and photographs. The planet appeared blue, due to its thick atmosphere of gases and a faint ring system of 13 main rings. Uranus was known to have five moons but a further ten, some less than 50 kilometres in diameter, were discovered by Voyager 2. A day on Uranus lasts for about $17\frac{1}{2}$ hours and a year is equivalent to 84 Earth years. Its largest moon is *Titania* with a diameter of 1600 km. All five moons are very cold and icy with a surface covered in craters and cracks. *Ariel* has deep wide valleys and *Miranda*, the smallest moon, is a mass of canyons and cracks with cliffs reaching up to 20km. Uranus has a greatly tilted axis so that some parts of the planet's surface are exposed to the Sun for half of the planet's orbit (about 40 years) and are then in continuous darkness for the rest of the time. Due to the tilt of the axis, the Sun is sometimes shining almost directly onto each of Uranus' poles during parts of its orbit.

urea an organic molecule, $CO(NH_2)_2$, that is a metabolic byproduct of the chemical breakdown of PROTEIN in mammals. In humans, 20-30 grams of urea are excreted daily in the urine, and although urea is not poisonous in itself, an excess of it in the blood points to a defective kidney, which will cause an excess of other, possibly poisonous, waste products.

V

vaccine a modified preparation of a VIRUS or BACTERIA that is no longer dangerous but can produce an immune response. It therefore gives immunity against infection with the actual disease. Vaccines can be administered by mouth or by a hypodermic syringe and are not effective immediately as it takes time for the recipient's IMMUNE SYSTEM to develop a memory for the modified virus or bacterium by producing specific ANTIBODIES.

vacuum in theory, a space in which there is no matter. However, a perfect vacuum is unobtainable and the term describes a gas at a very low pressure.

vacuum distillation distillation performed under reduced pressure. Since a reduction in pressure also reduces the boiling point of substances it means that certain substances which at normal pressures would decompose (break down), *can* be distilled.

vacuum tube *see* **diode.**

valency is the bonding potential or combining power of an atom or group, measured by the number of hydrogen ions (H^+, or equivalent) that the atoms could combine with or replace. In an ionic compound, the charge on each ion represents the valency, e.g., in NaCl, both Na^+ and Cl^- have a valency of one. In covalent compounds (see BONDS), the valency is represented by the number of bonds formed. In carbon dioxide, CO_2, carbon has a valency of 4 and oxygen 2.

  The *electronic theory of valency* explains bonds through the assumption that specific arrangements of outer electrons in atoms (outer shells of eight electrons) give stability (as with the inert gases, which have such a structure) through the transfer or sharing of electrons. Thus with the combination of sodium with chlorine, sodium has one electron in the outer shell, which it loses to chlorine to form the stable structure of the inert gas neon. Similarly, the gain of one electron by chlorine gives it the stable structure of argon.

valency electrons the electrons present in the outermost shell of an atom of an element and therefore the ones involved in forming bonds with other atoms when they are shared, lost or gained in the formation of a compound or ION. Some elements always have the same number of valence electrons, e.g., hydrogen has one, ordinary oxygen has two, and calcium has two.

valve in biology a piece of tissue attached to the wall of a tube that ensures the flow of blood is in one direction. The most important valves are the ones found in the HEART and VEINS, which prevent a backflow of blood. In engineering, valves are used in a similar way, to control the flow of fluids through pipes.

Van Allen radiation belts are two belts of radiation consisting of charged particles (electrons and protons) trapped in the Earth's magnetic field and forming two belts around the Earth. They were discovered in 1958 by an American physicist called James Van Allen. The lower belt occurs between 2000 and 5000km above the equator and its particles are derived from the Earth's atmosphere. The particles in the upper belt, at around 20,000km, are derived from the solar wind. The Van Allen belts are part of the Earth's magnetosphere, an area of space in which charged particles are affected by the Earth's magnetic field rather than that of the Sun.

Van der Graaf generator a machine that continuously separates electrostatic charges and in so doing produces a very high voltage. The fundamental structure of a Van der Graaf generator consists of a hollow metal sphere supported on an insulating tube. A motor-driven belt of, say, rubber or silk, carries positive charge from an electrode at the bottom of the belt into the sphere. The sphere gradually becomes positively charged, and in some generators of this type, voltages as high as 500kV or 10,000,000 volts can be produced. When used in conjunction with high voltage X-ray tubes, large machines with elaborate electrode systems can generate electrical energy, which is used to split atoms for research purposes.

Van der Waals' forces are weak, forces of attraction between two neighbouring atoms. They are named after the Dutch physicist, Johannes van der Waals (1837-1923), who first discovered their existence. In any atom, the electrons are continually moving and therefore have random distribution within the electron cloud of the atom. At any one moment, the electron cloud of an atom may be distorted so that a DIPOLE is temporarily produced. If two non-covalently bonded atoms are close enough together, the transient dipole in one atom will disturb the electron cloud of the other. This disturbance will create a similar dipole in the second atom, which will in turn attract the dipole in the first. It is the interaction between these transient dipoles that results in weak, non-specific Van der Waals' forces. Van der Waals' forces occur between all types of molecules, but they decrease in strength with increasing distance between the atoms or molecules.

vaporization see **evaporation**.

vapour pressure the pressure exerted by a vapour whether in a mixture of gases or by itself. Vapour pressure is a state of equilibrium between the vapour of a substance and its liquid form in a closed container. It will depend on temperature and the physical properties of the liquid.

variable a quantity that changes and can have different values, as opposed to a constant. In the equation $y = 3x^2 + 7$, x and y are variables, whilst 3 and 7 are constants. An INDEPENDENT VARIABLE is a variable in a function that determines the value of the other variable. A DEPENDENT VARIABLE has its value determined by other variables. So in this example, x is the independent variable and y is the dependent variable, because it depends on the value of x.

variable star a star with a brightness that varies with time, due in some circumstances to the star pulsating which affects surface temperature and size and thus the luminosity (brightness). Other stars show a variation because the brightness of their surface is not the same all over; or because the star is actually one of two stars in a *binary* system and when one passes in front of the other the brightness is reduced because the light from one is partly blocked.

vector any physical quantity that has both direction and magnitude. Vectors include displacement, velocity, acceleration and momentum. In biology, the term vector refers to the plasmid used to carry a DNA segment into the host's cells or the organism that acts as a mechanism for transmitting a parasitic disease, e.g. mosquitoes are vectors of malaria.

vegetative propagation a type of reproduction in which the non-sexual organs of the plant are capable of producing progeny. Vegetative propagation occurs naturally in certain plants, e.g., potato tubers and strawberry runners.

vein any thin-walled vessel that carries blood back from the body to the HEART. Veins contain few muscle fibres but have one-way VALVES that prevent backflow, thus enabling the blood to flow from body areas back to the heart.

In geology, a sheet-like feature usually occupying a fracture (crack) or fissure within a rock, which is infilled with mineral deposits. CALCITE and QUARTZ commonly form veins, but ore deposits do occur in this form, commonly mixed with other minerals.

velocity the rate of change of position of an object, that is the speed at which it travels. Velocity (v) is a VECTOR quantity, because it is expressed in both magnitude (size) and direction of travel. The unit of velocity is metres per second (ms^{-1}) and can be calculated using the displacement (s) and time elapsed (t) as follows: $V = s/t$. An object is described as moving with constant velocity when it is travelling along a straight line in equal proportions of distance against time. However, it is more likely that an object's velocity changes with time, in which case the object is said to be accelerating (see ACCELERATION).

velocity of light the VECTOR quantity for light travelling through a given medium. The velocity of light in a vacuum and in air hardly differs and is approximately $3.0 \times 10^8 ms^{-1}$. However, the velocity of light in water is approximately $2.3 \times 10^8 ms^{-1}$, and it is this difference in velocity in air and water that explains the REFRACTION of light when passing from one medium to another (*see also* SPEED OF LIGHT).

vena cava one of the two major veins that empty into the right chamber of the HEART. The superior vena cava (SVC) carries the blood collected from the upper part of the body, e.g. neck and brain, while the inferior vena cava (IVC) carries the blood from the lower half of the body, e.g., liver, kidney and legs.

ventricle a major chamber of the HEART, which is thick-walled and muscular as it is the main pumping chamber. The outflow of the right ventricle is known as the PULMONARY ARTERY, which takes blood to the lungs, and the outflow of the left ventricle is called the AORTA, which takes blood to the head and the rest of the body.

Venus is the second planet in the solar system with its orbit between those of Mercury and Earth and it is also the brightest. It is known as the Morning or Evening star. Venus is about 108.2 million km (67 million miles) from the Sun and is extremely hot with a surface temperature in the region of 470°C. It has a thick atmosphere of mainly carbon dioxide, sulphuric acid and other poisonous substances which obscure its surface. The size of Venus is similar to Earth with a diameter at the equator of 12,300 km. The atmosphere of carbon dioxide traps heat from the Sun (the greenhouse effect) allowing none to escape. Hence the surface rocks are boiling hot and winds whip through the atmosphere at speeds in excess of 320 km/hr. Venus is unusual in being the only planet to spin on its axis in the opposite direction to the path of its orbit. Also it spins very slowly so that a "day" on Venus is very long, equivalent to 243 Earth days. A year is 225 days. Venus has no satellites and because its surface is hidden, much of the known information about the planet has been obtained from SPACE PROBES. The *Magellan* space probe launched by the USA in 1989 visited Venus, sending back valuable photographs. *Venera 13*, a Russian probe landed on Venus in 1982, and obtained a rock sample and other information. The surface of the planet has been shown to be mountainous with peaks 12 kilometres high. It is covered with craters and also a rift valley. It is possible that there are active volcanoes.

vertex the point at which two sides of a polygon or the planes of a solid intersect.

virga a feature relating to certain CLOUD formations where trails of precipitation (rain or snow) fall beneath the cloud but evaporate before reaching the ground.

virus the smallest microbe, which is completely parasitic as it cannot grow or reproduce outside the cells of its host. Most, but not all, viruses cause disease in plant, animal and even bacterial cells. Viruses are classified according to their nucleic acids and can contain double-stranded (DS) or single-stranded (SS) DNA or RNA. In infection, any virus must first bind to the host cells and then penetrate to release the viral DNA or RNA. The viral DNA or RNA then takes control of the cell's metabolic machinery to replicate (multiply) itself, form new viruses, and then release the mature virus by either budding from the cell wall or rupturing and hence killing the cell. Some familiar examples of virus-induced diseases are herpes (double-stranded DNA), influenza (single-stranded RNA) and the retroviruses (single-stranded RNA, believed to cause AIDS and perhaps CANCER).

viscosity a property of fluids that indicates their resistance to flow. For example, oil is more viscous than water, and an object falling through oil is much slower than the same object falling through water because of the greater viscous force acting on it. A perfect fluid would be non-viscous.

vitamins a group of organic substances that are required in very small amounts in the human diet to maintain good health. A lack of a particular vitamin results in a deficiency disease. There are two groups of vitamins; those which are fat-soluble including A, D, E and K and those which are water-soluble, C (ascorbic acid) and B (Thiamine).

The six vitamin groups are as follows:

A or Retinol
Source: green vegetables, dairy produce, liver, fish oils
Needed for: the manufacture of rhodopsin—for night vision. Also for the maintenance of the skin and tissues
Deficiency: night blindness and possible total blindness

B complex including Thiamine, Riboflavin, Nicotinic acid, Pantothenic acid, Biotin, Folic acid, B_6, Pyroxidine, B_{12} (cyanocobalamin), Lipoic acid
Source: green vegetables, dairy produce, cereal, grains, eggs, liver, meat, nuts, potatoes, fish
Needed for: the manufacture of red blood cells, for enzyme activity and for amino acid production. Also important for nerves
Deficiency: Beri beri (B_1 deficiency), anaemia and deterioration of the nervous system (B_{12} deficiency)

C (ascorbic acid)
Source: citrus fruit, green vegetables
Needed for: maintaining the cell walls and the connective tissue. Aiding the absorption of iron by the body
Deficiency: scurvy— affects skin, blood vessels and tendons

D
Source: fish oils, eggs, dairy produce
Needed for: controls calcium levels required for bone growth and repair
Deficiency: rickets in children—deformation of the bones. Osteomalacia in adults—softening of bones

E
Source: cereal grains, eggs, green vegetables
Needed for: maintenance of cell membranes
Deficiency: unusual, as it is common in the diet

K
Source: leafy green vegetables, especially spinach, liver
Needed for: clotting of blood
Deficiency: rare as it is also manufactured by bacteria in the gut

viviparous a term describing any animal that gives birth to young that have developed inside its body. Viviparity is not restricted to mammals but also applies to some species of insect, e.g. the species of mite, *Acarophenox*, whose young develop by devouring and thus killing the mother.

volatile a term describing any substance that can easily change from the solid or liquid state to its vapour i.e. a substance with a high vapour pressure and which passes into the gaseous PHASE rapidly.

In geology, elements which would ordinarily be gaseous under normal conditions are dissolved in a MAGMA because of the pressure and nature of the melt. On reaching the surface, or reduced pressure, the volatiles turn back into gas, forming carbon dioxide, sulphur dioxide, hydrochloric acid and others. Volatiles in *coal* include methane and hydrogen and water vapour. Anthracite contains about 10% volatiles which increases to approximately 50% or more in peat.

volcanic bomb a lump of LAVA thrown out from a VOLCANO and which can take one of several shapes depending upon the type of lava, the degree of solidification and its flight through the air.

volcanic neck see **plug**.

volcanic vent the pipe which connects a volcanic crater with the source of MAGMA.

volcano a natural vent or opening at the Earth's surface, which is connected by a pipe or conduit to a chamber at depth that contains MAGMA. The eruptions may consist of steady flows of LAVA, explosions of ash, gas and fragments, or hot flows of ash. The type of eruption is determined by the amount of gas held in the lava and tends to be named after specific volcanoes or areas of volcanic activity e.g. Hawaiian (outpouring of fluid basaltic lavas), Peléean (violent eruptions with hot ash flows and viscous lavas), Strombolian (moderate eruptions with small explosions and basaltic lava of average VISCOSITY) and Vesuvian (very explosive after long dormant periods with frothy gas-filled lava and clouds of ash and gas). Volcanoes may be active, dormant or extinct. Dormant volcanoes have often been mistaken for extinct, only to erupt with surprising and often catastrophic results.

volt (V) the unit of POTENTIAL DIFFERENCE. One volt is equal to one joule per coulomb of charge, i.e. $V = JC^{-1}$.

voltage the electrical energy that moves charge around a CIRCUIT. Voltage is the same as POTENTIAL DIFFERENCE, is measured in VOLTS and is the potential energy given to every coulomb of charge from a battery.

volume the space occupied by any object or substance. The volume of a liquid will depend on the amount of container space it occupies, but the volume of any gas will vary with pressure and temperature. Volume is measured in cm^3 (cubic centimetres) or m^3 (cubic metres). The volume of a cube, cuboid or cylinder is equal to the area of the base x height; the volume of a pyramid or cone is equal to $1/3$ of the area of the base x height. The volume of a sphere is $4/3\pi r^3$.

volumetric analysis chemical analysis which uses standard solutions of known concentrations to calculate a particular constituent present in another solution, using TITRATION.

vulgar fraction an ordinary fraction with one number over the other, e.g. $3/5$, $5/9$, $1/16$.

vug a cavity in a rock which is often lined or infilled with crystals.

W

warm front the edge of a mass of warm air advancing and rising over cold air. Cloud develops in the rising air with heavy NIMBOSTRATUS forming as the front passes. There is usually heavy rain associated with the front. The temperature rises after the front with the rain clearing and often there is a change in wind direction.

water hydrogen oxide (H_2O), a compound that is found everywhere. It can occur as solid, liquid and gas phases. It forms a very large part of the Earth's surface and is vital to life. It occurs in all living organisms and has a remarkable combination of properties in its solvent capacity, chemical stability, thermal properties and abundance.

water table the level below which water saturates the available spaces in the ground. A spring or river is formed when, due to geological conditions, the water table rises above ground level. The position of the water table varies with the amount of rainfall, loss through evaporation and percolation through the soil.

watt (W) a unit of power that is the rate of WORK done at 1 JOULE per second, i.e. $1W = 1Js^{-1}$.

wave a mechanism of energy transfer through a medium. The origin of the wave is vibrating particles, which store and release energy while their mean position remains constant as it is only the wave that travels. Waves can be classified as being either LONGITUDINAL WAVES, e.g. sound, or TRANSVERSE WAVES, e.g., light, depending on the direction of their vibrations. There is a basic wave equation that relates the wavelength (λ), frequency (f), and speed (c) of the wave as $c = f\lambda$. All forms of waves have the following properties: diffraction; interference; reflection and refraction. ELECTROMAGNETIC WAVES have all of these properties but differ from ordinary waves, such as water waves, in that they can travel through a vacuum, e.g. outer space.

wavelength (λ) the distance between two similar and consecutive points on a wave, which have exactly the same displacement value from the rest position (that is, the same amplitude). An example would be the distance between two crests (maximum displacement) or two troughs (maximum displacement). Wavelength is a measure of distance and hence has units of metres (m).

weathering the action of physical and chemical processes at and just beneath the surface of the Earth which breaks down minerals and rocks. The action of frost widening cracks in rocks, moving water loosening particles and contributing to chemical reactions, and other actions all assist in the disintegration process.

weather map a graphic chart indicating measurements of temperature, wind speed, atmospheric pressure, cloud and precipitation which is used as a basis for forecasting.

weight the gravitational force of attraction exerted by the Earth on an object. The force of attraction exists between all objects, but it is small and the Earth's attraction is much larger. As weight is a FORCE, its unit is the NEWTON (N). The weight of any object on earth can be calculated using:

$$W = mg \text{ where } m = \text{mass (kg)}$$
$$g = \text{gravitational constant} = 9.8ms^{-2}$$

In everyday use, the term weight really refers to the mass of a person or object.

white blood cell see **leucocyte**.

white dwarf a type of star that is very dense with a low luminosity. They result from the explosion of stars that have used up their available hydrogen. Due to their small size, their surface temperatures are high and appear white (see SUPERNOVA).

white light light that contains all the visible wavelengths and which can therefore be broken down into a continuous spectrum of colours, e.g. as when shone through a prism.

work is done when an object is moved by a FORCE. A force is said to do work only when the object it is applied to moves in the direction of the force. The work done is calculated by multiplying the force (F) by the distance(s) through which it moves, i.e. W = FS. Work is measured in the unit of energy, the JOULE. For example, if you have to pull with a force of 50 newtons to move a box 3 metres in the direction of the force (toward yourself), then work = 50N x 3.0M = 150Nm = 150J.

Work is also the transfer of energy, that is the changing of energy into a different form, e.g. potential energy into kinetic energy, or chemical energy into heat energy.

X

x-axis the "horizontal" axis when constructing and plotting a graph.

X-chromosome one kind of CHROMOSOME that is involved in the determination of the sex of an individual. A woman has a pair of X-chromosomes, whereas a man has one X-chromosome and one Y-chromosome. There are many GENES on the X-chromosome which have nothing to do with the sex of the individual. For example, red-green colour blindness is determined by a RECESSIVE gene on the X-chromosome. If a woman carrying this gene has a son (X,Y) then he will inherit colour blindness as the Y-chromosome will have no corresponding gene to suppress (over-ride) the effect. If she has a daughter (X,X), and the father has normal vision, then the recessive gene is still inherited but its effect is suppressed (hidden) as the X-chromosome from the father will carry the dominant gene for normal vision.

xerography a copying process in which an ELECTROSTATIC image is formed on a surface when exposed to an optical image. A powder mix of GRAPHITE and a thermoplastic resin of opposite charge to the electrostatic image is dusted on to the surface and the particles cling to the charged areas. The image is then transferred to a sheet of paper, again through use of opposite charges, and the image is fixed by heat.

X-ray astronomy stars emit a variety of electromagnetic radiation including X-rays, but this is unable to penetrate the Earth's atmosphere to be detected. Very hot stars send out large amounts of X-rays which can be detected by equipment contained on space satellites. The satellite ROSAT launched in 1990 has equipment to undertake X-ray astronomy and obtain information about distant bodies in space (e.g. WHITE DWARFS and SUPERNOVAE) that are emitting X-rays.

X-ray crystallography (or X-ray diffraction— XRD – crystallography) a technique used in geology to identify minerals, and biology or chemistry to work out the structure of complex molecules. It involves directing a beam of X-rays at a CRYSTAL and the rays are diffracted (see DIFFRACTION) off the planes of atoms in the crystal. By repeating the procedure and then calculating the spacing between atomic planes, a representation of the crystal's structure can be determined.

X-ray fluorescence spectrometry (or XRF spectrometry) an analytical method in geology for determination of a wide range of elements in bulk rock specimens. Rock samples are prepared as ground powder compressed into flat cylinders or fused into coin-like flat discs and then excited with X-ray radiation. The radiation causes the removal of an ELECTRON from an ORBITAL, and when it is replaced, the surplus energy is emitted as a characteristic, secondary, X-ray. The X-ray is measured by the spectrometer and the intensity of the radiation compared to a standard to enable concentrations to be calculated. The technique is widely used in analysing rock samples both for major elements and certain TRACE ELEMENTS. Concentrations of trace elements as low as 1 to 10 parts per million can be detected although, in many instances, the quantities are higher.

Major Elements	Trace Elements
Na - sodium	Rb - rubidium
Mg - magnesium	Sr - strontium
Al - aluminium	Y - yttrium
Si - silicon	Nb - niobium
P - phosphorus	Zr - zirconium
K - potassium	Cr - chromium
Ca - calcium	Ni - nickel
Ti - titanium	Cu - copper
Mn - manganese	Zn - zinc
Fe - iron	Ga - gallium
	Ba - barium
	Pb - lead
	Th - thorium
	U - uranium

X-rays the part of the ELECTROMAGNETIC spectrum with a wavelength range of approximately 10^{-12} to 10^{-9}m and a frequency range of 10^{17} to 10^{21}Hz. X-rays are produced when electrons moving at high speed are absorbed by a target. The resultant waves will go through solids to varying degrees, depending upon the density of the solid. Hence X-rays of certain wavelengths will penetrate flesh, but not bone or other more dense materials. X-rays serve both therapeutic and diagnostic functions in medicine and are used in many areas of industry where inspection of hidden, inaccessible objects is necessary.

Y

y-axis the "vertical" axis when constructing and plotting a graph.

Y-chromosome the small chromosome that carries a dominant gene for male-ness. All normal males have 22 matched pairs of chromosomes and one un-matched pair, one large X-chromosome and one small Y-chromosome. The X-chromosome, which is inherited from the mother, carries many more genes than the Y-chromosome, which is inherited from the father. During sexual re-production, the mother must contribute one X-chromosome, but the father can contribute either an X or Y-chromosome. The effect of the Y-chromosome is that a testis develops in the embryo instead of an ovary. Thus the sex of the resulting offspring is dependent on the father's contribution—female (X, X) or male (X, Y).

yeast unicellular micro-organisms that form a fungus. Yeast cells can be circular or oval in shape and reproduce by spore formation. The enzymes secreted by yeasts are used in brewing and baking industries as they can convert sugars into alcohol and carbon dioxide.

yield point HOOKE'S LAW states that for a material such as steel, in wire form, the extension is proportional to the tension, up to what is called the elastic limit. An increase in tension beyond this limit takes the material to the yield point, where a sudden increase in elongation occurs with only a small further increase in tension.

Young's modulus Young's modulus (E) relates the STRESS and STRAIN in a solid (usually wire) using the following formula:

$$E = stress/strain = \sigma/\varepsilon$$

where σ = Force/Area and E = Change in Length/Length

Young's modulus has units of newtons per metre squared (Nm^{-2}) and is calculated only when the material is under elastic conditions, i.e., the force applied does not exceed the elastic limit and cause deformation.

Z

zenith the highest point over head—the pole which is vertically above the observer with the latter in the centre of the horizon circle. Zenith is opposite to the NADIR in the CELESTIAL SPHERE.

zenith distance the angular distance of a heavenly body from the ZENITH.

zeolites a group of alumina silicates of sodium, potassium, calcium and barium that contain loosely held water that can be removed by heating and regained by exposure to water. However, the cavities created by the loss of water can be occupied by other molecules of a similar size. Zeolites have thus found uses in ION EXCHANGE and as adsorbents. Zeolites containing small amounts of platinum or palladium are used as CATALYSTS in the CRACKING of HYDROCARBONS. There are both natural and synthetic zeolites.

zodiac a zone within the CELESTIAL SPHERE which contains the paths of the planet, the Moon and the Sun. It is divided into the signs of the zodiac, named after the constellations.

zoology a branch of biology that involves the study of animals. Subjects studied include anatomy, physiology, embryology, evolution, and the geographical distribution of animals.

zygote the cell produced by the fusion of male and female germ cells (GAMETES) during the initial stage of FERTILIZATION. The zygote is a DIPLOID cell, formed by the fusion of the haploid male gamete and the haploid female gamete.

zymogen a form of an ENZYME which is inactive. Most zymogens are inactive precursors (starting compounds) of enzymes in the pancreas, which are involved in protein digestion. Synthesis of these digestive enzymes as zymogens prevents the unwanted digestion of the tissue in which the enzyme was made. The zymogen becomes activated by chemical modifications to form its active version when it reaches its site of function, e.g., the enzyme chymotrypsin (digests protein) is synthesized in the pancreas as the zymogen, chymotrypsinogen, and becomes activated only when it reaches its destination, the small intestine.

zymurgy a branch of chemistry that involves the study of FERMENTATION processes.

Appendices

APPENDIX I

Periodic Table

Group	1 (IA)	2 (2A)	3 (3B)	4 (4B)	5 (5B)	6 (6B)	7 (7B)	8 (8B)	9 (8B)	10 (8B)	11 (1B)	12 (2B)	13 (3A)	14 (4A)	15 (5A)	16 (6A)	17 (7A)	18 (8A)
Period 1	1 H																	2 He
2	3 Li	4 Be											5 B	6 C	7 N	8 O	9 F	10 Ne
3	11 Na	12 Mg											13 Al	14 Si	15 P	16 S	17 Cl	18 Ar
4	19 K	20 Ca	21 Sc	22 Ti	23 V	24 Cr	25 Mn	26 Fe	27 Co	28 Ni	29 Cu	30 Zn	31 Ga	32 Ge	33 As	34 Se	35 Br	36 Kr
5	37 Rb	38 Sr	39 Y	40 Zr	41 Nb	42 Mo	43 Tc	44 Ru	45 Rh	46 Pd	47 Ag	48 Cd	49 In	50 Sn	51 Sb	52 Te	53 I	54 Xe
6	55 Cs	56 Ba	71 Lu *	72 Hf	73 Ta	74 W	75 Re	76 Os	77 Ir	78 Pt	79 Au	80 Hg	81 Tl	82 Pb	83 Bi	84 Po	85 At	86 Rn
7	87 Fr	88 Ra	103 Lr **	104 Rf	105 Db	106 Sg	107 Bh	108 Hs	109 Mt	110 Uun	111 Uuu	112 Uub						

* lanthanides	57 La	58 Ce	59 Pr	60 Nd	61 Pm	62 Sm	63 Eu	64 Gd	65 Tb	66 Dy	67 Ho	68 Er	69 Tm	70 Yb
** actinides	89 Ac	90 Th	91 Pa	92 U	93 Np	94 Pu	95 Am	96 Cm	97 Bk	98 Cf	99 Es	100 Fm	101 Md	102 No

APPENDIX 2

Element Table

Element	Symbol	Atomic Number	Relative Atomic Mass*
Actinium	Ac	89	{227}
Aluminium	Al	13	26.9815
Americium	Am	95	{243}
Antimony	Sb	51	121.75
Argon	Ar	18	39.948
Arsenic	As	33	74.9216
Astatine	At	85	{210}
Barium	Ba	56	137.34
Berkelium	Bk	97	{247}
Beryllium	Be	4	9.0122
Bismuth	Bi	83	208.98
Bohrium	Bh	107	
Boron	B	5	10.81
Bromine	Br	35	79.904
Cadmium	Cd	48	112.40
Caesium	Cs	55	132.905
Calcium	Ca	20	40.08
Californium	Cf	98	{251}
Carbon	C	6	12.011
Cerium	Ce	58	140.12
Chlorine	Cl	17	35.453
Chromium	Cr	24	51.996
Cobalt	Co	27	58.9332
Copper	Cu	29	63.546
Curium	Cm	96	{247}
Dubnium	Db	105	
Dysprosium	Dy	66	162.50
Einsteinium	Es	99	{254}
Erbium	Er	68	167.26
Europium	Eu	63	151.96
Fermium	Fm	100	{257}
Fluorine	F	9	18.9984

Francium	Fr	87	{223}
Gadolinium	Gd	64	157.25
Gallium	Ga	31	69.72
Germanium	Ge	32	72.59
Gold	Au	79	196.967
Hafnium	Hf	72	178.49
Hassium	Hs	108	
Helium	He	2	4.0026
Holmium	Ho	67	164.930
Hydrogen	H	1	1.00797
Indium	In	49	1114.82
Iodine	I	53	126.9044
Iridium	Ir	77	192.2
Iron	Fe	26	55.847
Krypton	Kr	36	83.80
Lanthanum	La	57	138.91
Lawrencium	Lr	103	{257}
Lead	Pb	82	207.19
Lithium	Li	3	6.939
Lutetium	Lu	71	174.97
Magnesium	Mg	12	24.305
Manganese	Mn	25	54.938
Meitnerium	Mt	109	
Mendelevium	Md	101	{259}
Mercury	Hg	80	200.59
Molybdenum	Mo	42	95.94
Neodymium	Nd	60	144.24
Neon	Ne	10	20.179
Neptunium	Np	93	{237}
Nickel	Ni	28	58.71
Niobium	Nb	41	92.906
Nitrogen	N	7	14.0067
Nobelium	No	102	{255}
Osmium	Os	76	190.2
Oxygen	O	8	15.9994
Palladium	Pd	46	106.4
Phosphorus	P	15	30.9738

Platinum	Pt	78	195.09
Plutonium	Pu	94	{244}
Polonium	Po	84	{209}
Potassium	K	19	39.102
Praseodymium	Pr	59	140.907
Promethium	Pm	61	{145}
Protactinium	Pa	91	{231}
Radium	Ra	88	{226}
Radon	Rn	86	{222}
Rhenium	Re	75	186.20
Rhodium	Rh	45	102.905
Rubidium	Rb	37	85.47
Ruthenium	Ru	44	101.07
Rutherfordium	Rf	104	
Samarium	Sm	62	150.35
Scandium	Sc	21	44.956
Seaborgium	Sg	106	
Selenium	Se	34	78.96
Silicon	Si	14	28.086
Silver	Ag	47	107.868
Sodium	Na	11	22.9898
Strontium	Sr	38	87.62
Sulphur	S	16	32.064
Tantalum	Ta	73	180.948
Technetium	Tc	43	{97}
Tellurium	Te	52	127.60
Terbium	Tb	65	158.924
Thallium	Tl	81	204.37
Thorium	Th	90	232.038
Thulium	Tm	69	168.934
Tin	Sn	50	118.69
Titanium	Ti	22	47.90
Tungsten	W	74	183.85
Ununbium	Uub	112	
Ununnilium	Uun	110	
Ununumium	Uuu	111	
Uraniun	U92	238.03	

Vanadium	V	23	50.942
Xenon	Xe	54	131.30
Ytterbium	Yb	70	173.04
Yttrium	Y	39	88.905
Zinc	Zn	30	65.37
Zirconium	Zr	40	91.22

*Values of the *Relative Atomic Mass* in brackets refer to the most stable, known, isotope.

Elements listed by symbol

Symbol	Element	Symbol	Element
Ac	Actinium	Mo	Molybdenum
Al	Aluminium	N	Nitrogen
Am	Americium	Na	Sodium
Ar	Argon	Nb	Niobium
As	Arsenic	Nd	Neodymium
At	Astatine	Ne	Neon
Au	Gold	Ni	Nickel
B	Boron	No	Nobelium
Ba	Barium	Np	Neptunium
Be	Beryllium	O	Oxygen
Bi	Bismuth	Os	Osmium
Bk	Berkelium	P	Phosphorous
Br	Bromine	Pa	Protactinium
C	Carbon	Pb	Lead
Ca	Calcium	Pd	Palladium
Cd	Cadmium	Pm	Promethium
Ce	Cerium	Po	Polonium
Cf	Californium	Pr	Praseodymium
Cl	Chlorine	Pt	Platinum
Cm	Curium	Pu	Plutionium
Co	Cobalt	Ra	Radium
Cr	Chromium	Rb	Rubidium
Cs	Caesium	Rh	Rhodium
Cu	Copper	Rn	Radon
Dy	Dysprosium	Ru	Ruthenium
Er	Erbium	S	Sulphur
Es	Einsteinium	Sb	Antimony
Eu	Europium	Sc	Scandium
F	Fluorine	Se	Selenium
Fe	Iron	Si	Silicon
Fm	Fermium	Sm	Samarium

Fr	Francium	Sn	Tin
Ga	Gallium	Sr	Strontium
Gd	Gadolinium	Ta	Tantalum
Ge	Germanium	Tn	Terbium
H	Hydrogen	Tc	Technetium
He	Helium	Te	Tellerium
Hf	Hafnium	Th	Thorium
Ho	Holmium	Ti	Titanium
I	Iodine	Tl	Thallium
n	Indium	Tm	Thulium
In	Iodine	U	Uranium
Ir	Iridium	V	Vanadium
Kr	Krypton	W	Tungsten
La	Lanthanum	Xe	Xenon
Li	Lithium	Y	Yttrium
Lu	Lutetium	Yb	Ytterbium
Md	Mendelevium	Zn	Zinc
Mg	Magnesium	Zr	Zirconium
Mn	Manganese		

The names and symbols of elements 104 to 112 in this table have been approved by the International Union of Pure and Applied Chemistry (IUPAC). There are a further six symbols for elements that can reliably be predicted but have not, as yet, been discovered. These are as follows:

113	Uut
114	Uuq
115	Uup
116	Uuh
117	Uus
118	Uuo

All these elements exist for a very short time in laboratory experiments and decay radioactively.

APPENDIX 3

The Greek Alphabet

Name	Capital	Lower Case	English Sound
alpha	A	α	a
beta	B	β	b
gamma	Γ	γ	g
delta	Δ	δ	d
epsilon	E	ε	e
zeta	Z	ζ	z
eta	H	η	e
theta	Θ	θ	th
iota	I	ι	i
kappa	K	κ	k
lambda	Λ	λ	l
mu	M	μ	m
nu	N	ν	n
xi	Ξ	ξ	x
omicron	O	o	o
pi	Π	π	p
rho	P	ρ	r
sigma	Σ	σ	s
tau	T	τ	t
upsilon	Y	υ	u
phi	Φ	φ	ph
chi	X	χ	kh
psi	Ψ	ψ	ps
omega	Ω	ω	o

APPENDIX 4

The International System of Units (SI units)

Quantity	Symbol	Unit	Symbols
acceleration	a	metres per second squared	ms-2 or m/s2
area	A	square metre	m2
capacitance	C	farad	F(1F = 1 AsV-1)
charge	Q	coulomb	C(1C = 1 As)
current	I	ampere	A
density	ρ	kilograms per cubic metre	kgm-3 or kg/m3
force	F	newton	N(1N = 1 kg ms-2)
frequency	f	herz	Hz(1Hz = 1s-1)
length	I	metre	m
mass	m	kilogram	kg
potential difference	V	volt	V(1V = 1JC-1 or WA-1)
power	P	watt	W(1W = 1Js-1)
resistance	R	ohm	Ω (1Ω = 1VA-1)
specific heat capacity	c	joules per kilogram kelvin	Jkg-1 K-1
temperature	T	kelvin	L
time	t	second	s
volume	V	cubic metre	m3
velocity	v	metres per second	ms-1 or m/s
wavelength	λ	metre	m
work, energy	W, E	joule	J(1J = 1Nm)

APPENDIX 4 (cont.)

Useful prefixes adopted with SI units

Prefix	Symbol	Factor
tera	T	10^{12}
giga	G	10^{9}
mega	M	10^{6}
kelo	k	10^{3}
hecto	h	10^{2}
deda	da	10^{1}
deci	d	10^{-1}
centi	c	10^{-2}
milli	m	10^{-3}
micro	μ	10^{-6}
nano	n	10^{-9}
pico	p	10^{-12}
femto	f	10^{-15}
atto	a	10^{-18}

APPENDIX 5

Geological Time Scale

Eon	Era	Sub-era	Period	Epoch	Millions of years since the start
PHANEROZOIC	Cenozoic	Quaternary	Pleistogene	Holocene	0.01
				Pleistocene	2.0
		Tertiary	Neogene	Pliocene	5.1
				Miocene	24.6
			Palaeogene	Oligicene	38
				Eocene	55
				Palaeocene	65
	Mesozoic		Cretaceous		144
			Jurassic		213
			Triassic		248

APPENDIX 5 (cont.)

Geological Time Scale

Eon	Era	Sub-era	Period	Epoch	Millions of years since the start
PHANEROZOIC	Palaeozoic	Upper Palaeozoic	Permian		286
			Caroniferous		360
			Devonian		408
		Lower Palaeozoic	Silurian		438
			Ordovician		505
			Cambrian		590
PROTEROZOIC	P R E C A M B R I A N				2500
ARCHAEAN					4000
PRISCOAN					4600

APPENDIX 6

The Solar System

Planet	Diameter at the Equator km	Mass relative to the Earth[1]	Average distance from Sun km[6]	The planet's "year"
Mercury	44840	0.054	57.91	87.969 days
Venus	12300	0.8150	109.21	224.701 days
Earth	12756	1.000	149.60	365.256 days
Mars	6790	0.107	227.94	686.980 days
Jupiter	142700	317.89	778.34	11.86 years
Saturn	120800	95.14	1427.01	29.46 years
Uranus	50800	14.52	2869.6	84.0 years
Neptune	48600	17.46	4496.7	164.8 years
Pluto	3500	0.1 (approx)	5907	248.4 years
Sun	1392000	332.958		
Moon	3476	0.0123		

[1] The mass of the Earth is 5.976×10^{24} kg